AT

Assembly Automation

A MANAGEMENT HANDBOOK

Second Edition

Assembly Automation

A MANAGEMENT HANDBOOK

Second Edition

FRANK J. RILEY

Industrial Press Inc.

Library of Congress Cataloging in Publication Data

Riley, Frank J.
    Assembly automation : a management handbook / Frank J. Riley. —
2nd ed.
    Includes index.
    ISBN 0-8311-3041-5
    1. Assembling machines. 2. Production management. I. Title
TJ1317.R54 1996
670.42'7 — dc20

 # 35 2 35155
1885198

96-33563
CIP

INDUSTRIAL PRESS INC.
200 Madison Avenue
New York, New York 10016-4078

Second Edition, 1996

ASSEMBLY AUTOMATION

First Printing

10 9 8 7 6 5 4 3 2 1

Contents

CHAPTER 13. AUTOMATED ASSEMBLY IN THE 21st CENTURY

PREFACE TO THE SECOND EDITION

This second edition of *Assembly Automation*, coming six printings and a decade after the publication of the first edition, reconfirms the author's belief and experience that selection and implementation of major factory automation on an economically rewarding basis is the most serious career responsibility that most manufacturing managers or engineers will ever face. With few exceptions the opportunity to select and implement such major systems occurs only a few times in one's corporate employment, and success or failure will determine one's future career. It will also greatly impact on the future course and success of any division of the manufacturing organization itself.

The environment in which manufacturing managers and engineers work as we approach the twenty-first century is radically different from the high technology euphoria of the early 1980's in which the first edition was written.

Management's former emphasis on emerging technology tools of the early 1980's as a panacea for global competitiveness has fallen to a far more conservative posture, a more realistic expectation of what can be done with appropriate levels of capital investment.

Quality has emerged as the single most dominant factor determining automation justification and implementation. Flexibility or agility, in the sense of provision for continuous improvement of processes and product changes, is another critical justification factor. At the same time, reengineering or staff downsizing has reduced internal automation development capabilities, increasing corporate dependence on outside supplies for automated system development.

The significant acceptance of the first edition was due in large part to its use as a textbook for management level seminars in some sixteen countries and throughout the United States.

The author and both professional and academic colleagues have had the opportunity, in leading such seminars, to evaluate the organization of the topics in this book and the relative importance of various sections of the first edition. This edition takes these factors into account.

This second edition has also taken into account significant changes in management culture. It also owes much to a series of major international studies examining future factory automation in a global environment.

Automated assembly—both conventional and robotic—over the past forty years has been widely discussed in seminars, workshops, and conferences and even more broadly covered in trade journals and academic papers. Strangely enough it still has produced little permanent literature and much of that has had very restricted practical application because of its highly theoretical content. Other works in this field are how-to-do-it oriented without examining the whys and whereas that concern the management of most potential users of automatic assembly.

In this book, there has been a basic assumption that real significant growth in use of automatic assembly, as in any emerging technology, will come from the wise selection and proper application of commercially available, field ready, equipment. The book is intended to be a practical guide in this selection and application process.

Automatic assembly, as discussed in this book, is the repetitive assembly of manufactured products by use of programmed or sequentially controlled mechanisms to obtain a variety of management objectives such as reduced inventories, process flexibility, reduced time-to-market, higher product quality, and which may or may not include reduced direct labor costs.

For many reasons, it is easier to state what the book is not intended to be than to state briefly what areas it does cover. Automatic assembly is usually not true automation. If one accepts the classical definition of automation which includes control feedback, most assembly equipment has a variety of functional options available when assembly or product functionality failure is detected but usually these options do not normally, for the vast majority of systems, include fully automatic correction or repair of an improper assembly itself. There is a determined effort at this time to determine the extent of both technical and financial feasibility for such capability. This inability to correct a deficient assembly *automatically* or to automatically restore defective operations within specific control limits separates most automatic assembly from true automation. Some recent systems, mostly at customer insistence, have attempted such automated corrective action. These are difficult to implement and require an unusually high level of competency on the part of the user workforce.

Automatic assembly is often joined to packaging, but is not packaging. Packaging involves placing finished products or goods in containers for purposes of storage, identification, marketing, and dispensing. The product container, having served this need, is no longer useful and is discarded.

Automatic assembly *is* involved with placing discrete parts in specific spatial relationships within housings, bases or cases which must be retained if the product is to function. In real life there is too little synergistic communication between packaging and assembly machine manufacturers. Construction and control of equipment design differs widely in these two related and interdependent industries.

The study of automatic assembly is not a course in ingenious mechanisms or machine design, since mechanisms are only part of the effort needed for successful mechanized assembly. The publishers of this book offer an excellent multivolume series, *Ingenious Mechanisms for Designers and Inventors* edited by Franklin Jones, et al., which is most useful for machine designers.

Dr. Bruno Lotter has written a most excellent text, *"Manufacturing Assembly Handbook"* (Butterworth & Co. Ltd.-UK) which is complementary to the contents of this book.

If automatic assembly is usually neither automation nor packaging nor mechanism design, what then is it?

This book is intended to show that successful automatic assembly is *an integrated systems approach* to a reduction in the largest areas of direct and indirect manufacturing costs in product assembly.

A great deal of justification for automatic assembly lies in its ability to meet today's sociological problems of broad economic swings, perceived declining rates of productivity, global manufacturing competitiveness and increasingly aggressive consumerism with its emphasis on absolute product quality. This justification process is not only a question of how, but why, when, where, and if one should mechanize assembly processes. The following chapters are quite specific that many proposed automated assembly processes do not lend themselves to significant mechanical assembly techniques, not only for technical reasons but because of failure to address major management concerns, or because senior corporate management is more interested in finance than manufacturing.

There is little in this book that is purely theoretical or purely academic. What follows is a systematic statement of decades of experiences and observations in the selection, procurement, design, building, and installation of automatic assembly systems. The book does not ignore theory, however. Many of today's practical problems of automated assembly implementation result from ignorance of basic underlying physical and mathematical principles which directly affect assembly machine productivity and economic success or failure.

In many ways this book may have greater value in helping its readers avoid assembly system failure, absolute or chronic, than in providing all of the input necessary for a successful system. All successful assembly systems contain some small but vital unique element of creativeness. This unique component may be in tooling development; it may be in new or revised ways of packaging or bundling existing or previously utilized technology; it may not be in the system design at all but in slight product design or process modifications, or even in being particularly sensitive to the customs, attitudes and skills of the equipment operators.

The creativity necessary to come up with unique production solutions can rarely be taught. Creativity alone, however, is not sufficient to produce a successful assembly system. Different is not necessarily better. Uniqueness must be based on a foundation of both successful experience and scientifically derived principles.

This book, then, is intended to share a lifetime of experience in assembly mechanization and define those principles contained in a multitude of case histories which these experiences have suggested.

As in the first edition, there is no section on electronic assembly. This is not an idle decision. In acting as author and editor of *The Electronics Assembly Handbook* (IFS-Springer-Verlag 1988) it was clear that the problems of electronic assembly are quite different. Electronic component suppliers standardizing on packaging in bandoliers, film carriers and magazine tubes have led to fully standardized electronics assembly systems, with little customizing done by major equipment producers.

Robots are considered in this book as electromechanical programmable transfer devices.

In an increasingly competitive world where manufacturing proficiency is vital to national strength, and where time poorly spent is money lost, it is a lot faster and far cheaper to learn by history than by personal involvement. For those involved in the selection, procurement, design, approval, and installation of assembly systems, this book is intended to outline a systematic approach to putting a successful assembly line on stream. As in any book of this type there are going to be judgment values stated. It is hoped that these are as objective as possible. If there is any bias in the statement of these values, it is going to be on the side of economic rather than technical feasibility, on what is financially practical, rather than what is theoretically possible. It will tend to oversimplify in order to illustrate principles, to show things in terms of black and white. The reader can find his or her own pleasure in discovering the many hidden shades of gray.

Frank J. Riley

Chapter 1
Why Automatic Assembly

INTRODUCTION

Automatic assembly is an optional capital investment on the part of management. Many areas of manufacturing require the purchase (or rental) of fabricating machines, such as presses, molding machines, and other parts fabricating equipment. A management decision to manufacture a product in high volumes mandates the purchase or rental from suppliers of fabricating equipment. This is not true when assembly equipment is considered. The availability of relatively inexpensive and unskilled labor is an optional management choice for most types of assembly work, and this choice may require little or no capital investment.

As production requirements go up and labor costs increase, automatic assembly becomes more attractive, in terms of direct labor cost reduction, as a sound capital investment. In the past most management did not go further than reviewing the direct labor savings in evaluating this potential investment.

The full profit potential in utilizing mechanical assembly, however, will be found in several other areas of equal or greater interest to management; indirect labor costs, quality with all of its warranty and liability implications, market

considerations, time-to-market, inventory reduction, and operator safety are all areas of potentially increased profit through the use of mechanized assembly.

Automatic assembly is a *high-production* tool. It is relatively expensive and will usually involve some degree of investment risk. Its broadest applications will come where production of products such as automotive components, electrical switches, etc., is measured in millions of annual units of production. These volumes may be, and usually are, made up of a family of similar products, particularly where the sequence of assembly is common. (Figure 1-1.)

Larger products, such as cars and larger appliances, may not need such high rates of annual production for economic justification. Products with high-quality requirements and products with seasonal demand may justify mechanized assembly on lower volumes.

Automatic assembly on any major scale must be a systems approach to manufacturing. Unless it is thought of as a systems approach, there will be little recognition of the demands and discipline automated assembly will impose on a manufacturing operation and all that this implies. It also means that unless it is viewed as a new approach to manufacturing, the enormous capabil-

Fig. 1-1. Typical door hardware assemblies. Rectangular and cylindrical faceplates, a variety of faceplate knob and rose shapes and finishes, and two standard lengths are accommodated in two assembly systems to improve justification.

ities it holds for improved marketing, quality control, and work scheduling will go unrecognized.

This chapter is aimed at a fuller understanding of the reasons for considering capital investment in mechanized assembly equipment. Without a good answer to why automatic assembly, the hows, whens, and wheres are not of importance.

THE ECONOMIC BASIS FOR AUTOMATIC ASSEMBLY

Some years ago the late Felix Giordano, Editor of SME's *Manufacturing Engineering*, wrote an editorial, "Will It Make Money?" Its brief message is basic to this chapter. In part it said:

"Manufacturing engineers bear the responsibility of selecting manufacturing equipment. For it is they who make the recommendations to buy this or that type of machine. It is they who write the specifications and recommend prospective sources. On those specifications and on those recommendations they often stake their own future with the organization.

So they have two functions. The first is to select the type of machinery that best suits the needs of their anticipated production. There is

no absolutely best machine, but there is the best one—or two or three or even four—for a given task under a given set of conditions. Only he or she knows the special needs and conditions that exist in their plant, and those that are likely to exist during the expected life of the machine they are to buy.

Once he makes the decision, it is up to him to make sure that the machine performs as expected. From this point on, he and the machine manufacturer are allies.

But what is the machine supposed to do? Ask most engineers and they will say that the machine must mill, or turn, or grind, or assemble. They will add such adverbs as well, fast, efficiently, repeatedly.

That is all very fine. But most of all the machine must make money. That is the name of the business—any business. To put it another way, the machine must give the most returns for the least investment.

Selection of equipment is a management function of the manufacturing engineer. The language of management is money.

Of course the machine must mill, or turn or assemble. Of course the engineer must ask whether a given machine has the capability which he knows he is going to use. But, in the end, the engineer must ask: Will this machine make money for my company? If he or she is really on the ball they will ask: Will this machine make the most money?"

Automatic assembly is a tool available to high-volume manufacturers to reduce the two most significant expense areas, product assembly and product quality. This potential can be fully realized only if there is a full understanding of the areas of economic justification for mechanized assembly system procurement.

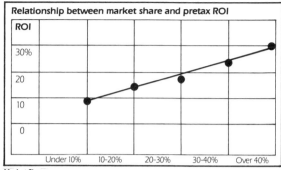

Fig. 1-2. Relationship between market share and pretax return on investment.

Market Share

If profit is the name of the game, then market share is an important consideration. In an article in the *Harvard Business Review* entitled "Market Share: Key to Profitability" it was stated:

"It is now widely recognized that one of the main determinants of business profitability is market share. Under most circumstances, enterprises that have achieved a high share of the markets they serve are considerably more profitable than their smaller-share rivals. This connection between market share and profitability has been recognized by corporate executives and consultants, and it is clearly demonstrated in the results of a project undertaken by the Marketing Science Institute on the Profit Impact of Market Strategies (PIMS). The PIMS project is aimed at identifying and measuring the major determinants of return on investment (ROI) in individual businesses. Phase II of the PIMS project reveals 37 key profit influences, of which one of the most important is market share.

There is no doubt that market share and return on investment are strongly related. Figure 1-2 shows average pretax ROI figures for groups of businesses in the PIMS project that have successively increasing share of their markets. On the average, a difference of 10 percentage points in market share is accompanied by a difference of about 5 points in pretax ROI.

Granted that high rates of return usually accompany high market share, it is useful to explore the relationship further. Why is market share profitable? What are the observed differences between low and high-share businesses? Does the notion vary from industry to industry? What does the profitability/market-share relationship imply for strategic planning? In this article we shall attempt to provide partial answers to these questions by presenting evidence on the nature, importance, and implications of the links between market share and profit performance.

The data reveal four important differences between high-share businesses and those with smaller shares. The samples used are sufficiently large and balanced to ensure that the differences between them are associated primarily with variations in market share, and not with other factors. These differences are:

As market share rises, turnover on investment rises only somewhat, but profit margin on sales increases sharply. ROI is, of course, dependent on both the rate of net profit on sales and the amount of investment required to support a given volume of sales. The ratio of investment to sales declines only slightly, and irregularly, with increased market share. The data show too that capacity utilization is not systematically related to market share.

The biggest single difference in costs, as related to market share, is in the purchases-to-sales ratio. In large-share businesses—those with shares over 40%—purchases represent only 33% of sales, compared with 45% for businesses with shares under 10%.

How can we explain the decline in the ratio of purchases to sales as share goes up? One possibility, as mentioned earlier, is that high-share businesses tend to be more vertically integrated, they "make" rather than buy, and often they own their own distribution facilities. A low purchases-to-sales ratio goes hand in hand with a high level of vertical integration.

Other things being equal, a greater extent of vertical integration ought to result in a rising level of manufacturing costs. But the data show little or no connection between manufacturing expense, as a percentage of sales, and market share. This could be because, despite the increase in vertical integration, costs are offset by increased efficiency."

How does a company gain market share? Advertising, product superiority, recognition of the market demands, all of these are vital, but, basically, market share falls to the determined, to those companies who can supply large quantities of high-quality, attractive merchandise at competitive prices. Some years ago market share belonged to the skilled producers: American, German, and English firms who enjoyed the availability of skilled, motivated workforces. Changing social patterns and emerging industrial powers such as Japan then upset traditional market patterns. Today, older industrial powers struggling to retain market share are often crippled by obsolete or obsolescent facilities, declining pools of skilled labor, falling productivity and a lack of motivation, work ethics, patriotism, and pride in their workforces, while emerging countries using world-class production systems, compete effectively in a global economy.

This means that, for mature industrial countries, in addition to attempting to reeducate and if possible to remotivate workforces to produce more efficiently, management must provide the necessary tools: appropriate levels of capital equipment to produce "more, better, for less."

For that reason, traditional capital justification formulas must be replaced by a closer examination of the indirect cost reduction areas offered by mechanical assembly systems and by the probability of improved quality and improved delivery offered by the wise use of such assembly systems.

Cost Reduction

There are many areas of cost reduction that can be utilized in justification for assembly machinery. Direct labor reduction is a most important consideration, but not necessarily the main reason for justification. Reduction of indirect labor, such as lead men and women, area supervisors, payroll personnel clerks and inspectors, is a potential area of indirect labor savings. Additionally, well planned mechanized assembly should substantially reduce work-in-process inventories. Because of substantial productive capacity through assembly automation, manufacturers can keep completed parts inventory to the lowest reasonable levels. Additionally, there are cost savings to be realized through the uniformity of assembly quality, a reduction in possible warranty or product liability claims, and the unique tendency of automatic assembly to increase overall plant efficiency. Examine each of these considerations in detail. All of them can be substantial factors in justifying mechanized assembly. Increasingly, indirect cost reduction aspects may be of far more current interest to corporate management than the direct labor savings.

Direct Labor Cost Reduction. Many older engineers still consider present direct labor costs as the prime area for justifying mechanized systems. If the engineer goes only this far, and does not include such factors as probable increases in payroll costs and continuing declines in productivity, justification becomes a study in past history, instead of projected actual labor costs.

With highly cyclical employment, there are other considerations today in the costs incurred when hiring and laying off personnel. Many states attribute direct costs of unemployment compensation to specific corporate experience. The cost of guaranteed annual wages is another indirect cost. Each person hired in the assembly area creates a potential liability for future operating costs, should it become impossible to maintain their employment on a permanent basis. On the other hand, automatic assembly equipment can be turned off and incur no additional costs other than the depreciation expenses in the first years of production incurred during a period of idleness. Again, cyclical employment involves training of new workers and retraining of reassigned workers, with substantial cost penalties in learning curves and in assembly quality. Mechanical assembly systems can avoid a great deal of these expenses by substantially reducing the amount of personnel turnover. (Figure 1-3.)

When production requirements are cyclical in nature, such as requirements for items sold on a seasonal basis (for example, outboard motors or lawn mowers), mechanized assembly allows maximum seasonal production capability with a minimum of restaffing. Plants may be faced with exceptional sales opportunities if fast delivery can be guaranteed and accomplished. Again, placing assembly machines on additional shift operation can easily meet production surges with very modest additional training or hiring.

The problems of seasonal hiring and unemployment caused by seasonal economic cycles is compounded when the hiring and discharge sequence is torn between seniority considerations, civil rights programs, and the company's need to retain particular skills. It is in the very nature of automatic assembly to stabilize employment levels.

Indirect Labor Reduction. Reduction through successful assembly automation in required floor supervisors, inspectors, and support personnel, such as payroll, personnel, food handling and health personnel, can be substantial. In one instance, a successful battery of ordnance assembly machines effected a reduction of 56 people in

Fig. 1-3. A modern assembly system. Two linear assembly machines with a spring winder, rotary assembly machine and conveyors make up a total assembly system.

direct labor, but also reduced the required number of supervisory, quality control, and maintenance people by 62 people. Since many of these costs are buried in factory burden, they may be more difficult to isolate, but these are truly savings. Overhead personnel incur the same type of fringe hiring and training costs as direct laborers.

One of the most frustrating problems at time of justification is to determine the new overhead costs when there is substantial reduction in direct labor forces in a given department. The overhead ratios will change drastically, and it may be necessary to add new technical skills in a department that was essentially people oriented. Failure to realize and state this may vitiate an excellent cost reduction program.

Product Liability and Warranty Costs. The use of automatic assembly equipment has a unique tendency to force upgrading of overall product quality. Its efficient use requires the maintenance of uniform levels of quality in assembly components by the very fact that wide variations in components quality will inhibit the productivity of the machinery. However, there are broad capabilities for the assurance of superior assembly quality in such machines beyond this fact. It is possible and most inexpensive to inspect for the presence, position, attitude, and orientation of each of the component part pieces during the developing assembly process itself. It is also possible to inspect for such functional characteristics as capacitance, impedance, resistance, vacuum decay or leak rate, dimensional characteristics, and torque. Without any additional labor content, 100% inspection can be realized, data recorded, and the product so coded as to be able to document that care in production required of prudent men and women by current liability legislation and judication. A large number of a such inspection stations are usually moderate in cost. Parts are already staged in fixtures. Fault counters can give objective data regarding quality trends in component parts. These data are extremely useful in adjusting prior fabricating and forming operations to required quality levels, especially where products are made in a work cell environment.

One most dramatic example for inspection expenditure occurred when a leading Japanese car manufacturer ordered a brake cylinder assembly system. Japanese project engineers objected to the cost of automatic testing stations. The general manager of the Japanese firm responded to their concern by asking what the postage might be for registered mail to recall automobiles with defective brakes let alone the impact on their market image. The cost of the necessary inspection functions was minor in comparison with the possible risk.

Emphasis on the uniformity of quality, the ability to document quality, and the ability to inspect 100% of all production should be of substantial interest to those concerned with the marketing and legal aspects of the corporate operation. Such so-called intangibles tend to be excellent positive factors of an overall justification request.

Production and Inventory Control Consideration. Automatic assembly systems are highly productive where volume requirements are high. Multishift operation is often indicated by the high capital expenditures for assembly systems. The ability of mechanized assembly systems to respond immediately without incurring learning curves to marketing opportunities give salespeople in the field the advantage of being able to accept orders on the basis of quick delivery or high volume knowing that the capabilities exist within the plant to meet these goals. This can be done without the normal "ramp-up" delays incurred in rehiring, restaffing, and retraining assembly departments. Governmental directives concerning such cyclical hiring and training have become so restrictive that, in many instances, operating managers are reluctant to do hiring on a short-term basis. Possession of automatic assembly equipment means that the new production goals can be achieved with a minimum of hiring or rehiring.

Additionally, work-in-process can be dramatically reduced. In one instance, conversion from solvent bonding to ultrasonic welding of a small plastic pump assembly permitted work-in-process times to be reduced from approximately 6 days to 45 seconds. It was possible to eliminate the in-process storage of several hundred thousand incomplete assemblies during their curing and material-handling cycles. Again, since machines can quickly respond to production goals, completed parts assembly can be kept to absolute practical minimum. (Figure 1-4.)

The limitations of assembly equipment, however, present counterbalancing requirements. Sufficient quantities of all necessary components must be available to the machine at any time production is required. This involves strong inventory controls and smooth material-handling operations to ensure uniform work flow to the machine itself.

Marketing Considerations

Knowledge that mechanized assembly systems are used to produce a product can give the marketing department significant psychological support and increased flexibility in the marketplace. It is difficult to be a good salesperson if you have doubts as to your product's quality, reliability, price, and delivery. Successful assembly systems provide psychological reassurance and competitive advantages in the marketplace.

Lowest Unit Cost. Successful mechanized assembly has, at least in theory, reduced assembly costs to their lowest level or lowest unit assembly cost. Once this is achieved, low-cost labor in other areas of the world no longer offer any labor cost advantage. Because of the low labor content and reduced factory size with mechanized assembly, it is often possible to place the assembly system in close proximity to major markets, reducing pipeline cost and delivery time. Manual assembly, on the other hand, is often done in areas remote from major market areas. Knowledge that production costs are minimal and equal to or below competitors' assembly costs may have

Fig. 1-4. A plastic pump assembly machine. When assembled by hand and using solvent bonding, several hundred thousand pumps were somewhere on the assembly line. Conversion to mechanical assembly and ultrasonic welding reduced parts in process to less than fifty.

significant impact on pricing strategy, and, where legal, provide for selective pricing.

Improved Deliveries. Mechanized assembly systems can offer better deliveries in several ways. Some firms enjoy stable production rates. Most plants, however, find required production rates varying on a weekly or daily basis. The problem is compounded in that major component producers manufacturing in annual volumes of millions of units often are required to produce products to small batch size order with a variety of customer-specified options. Mean order size for such customer-specified products (with cumulative annual production in millions of units) may be in quantities as small as 100 or less units. Finished goods inventory is, by necessity, restricted, so that production must follow receipt of specific orders and customer specifications.

Mechanized assembly systems can translate component inventories into finished products in a minimum of time and remain idle during slack periods without the necessity of finding, training, or reassigning a large number of people.

When quality assurance is done as part of the production function of an automated assembly system, it is not necessary to delay delivery of the assembly for inspection prior to shipment.

Mechanized assembly systems are not only rapid in their assembly function, but they eliminate the delays and probable quality deterioration resulting from material handling between single station or manual assembly processes. Assembly lead times are dramatically shortened.

When production demands are seasonal—for example, lawn mower production—or have short economic cyclic demands, relatively modest annual requirements (e.g., 400,000 or 500,000 units per year) must be produced at an annualized rate of several million units a year. The productivity of most good assembly systems means that assembly and hence deliveries can be made as close to time of sale as possible, without the need to use costly seasonal labor. Costs of inventory are drastically reduced.

Uniform Quality. In an age of consumerism, product quality is of paramount concern to the marketing department. It is an axiom in marketing that editorial comment has much greater

readership than commercial advertising and is considered (whether justified or not) to be more truthful. Enormously expensive marketing and advertising campaigns have been completely undermined by reports in the media of product defects or product recalls instituted by the manufacturers or by governmental edict.

Mechanized assembly with its capability of inspecting 100% of the assemblies produced and, if necessary, to document and even codify individual units of production offers the greatest assurance that recalls, field service problems, and warranty expenses with their negative impact on sales are held to an absolute minimum.

In addition, the use of mechanized assembly forces uniform product quality, a consistency of quality that may be relied on both by salespeople and customers.

Strategic Pricing. Significant cost reductions achieved by mechanized assembly may be used for strategic pricing considerations. Most high-volume-production plants find that a few basic products produce a major part of their revenue, while a multitude of low volume products consume a disproportionate share of cost and attention. Such low volume products are justified as being necessary by the marketing department to fill out the product line. It is usually difficult to place full burden on such low volume products, and hence their manufacturing cost is often underwritten by more profitable products. Increased profitability in major product lines through assembly mechanization can be used to continue this subsidy or even increase the subsidy temporarily on low volume products to the point that even these products can become priced competitively with the hope of increasing market share.

Excess capacity on a machine may be used to secure new marketplaces. One major industrial component maker enjoyed a large market share and high profitability on a product: profitability that was enhanced by the installation of a successful assembly machine. This company had never been successful because of its pricing structure in government sales. Since the machine had substantial reserve capacity and its payback was fully justified on existing industrial sales, it was decided to quote on a GSA request using material, variable costs, and profit only as the basis for the price offered the government. Profit on a single major government supply contract *utilizing the machine's reserve capacity returned the entire purchase price of the machine.*

Sociological Considerations

The ability of a mechanized assembly system to verify the presence and position of each component and the functionality of the assembly as it progresses, and the additional capability to quantify and record objectively failure modes and if necessary to codify or serialize each unit of production, offers major benefits to manufacturers operating in a hostile global economy. These benefits alone are often sufficient to justify procurement of an assembly system. Not all these benefits are intangible. Some are very specific cost reduction items; others are insurance against fiscal disaster.

Reduced Warranty Expenses. Many products such as automotive components that could be assembled automatically are often incorporated into larger, more expensive assemblies. Typically these items may be switches, solenoids, pumps, small motors, relays, and other mechanical, fluid power, or electrical devices costing three or four dollars. Once combined into an automobile, dish washer, air conditioner, or other major assembly they are expected to function reliably for years. If they fail, field service may be required at a cost many, many times the purchase cost of the original subassembly.

At one time this replacement cost was picked up by the product user. Increasingly, service and replacement costs are borne by the producer of that subassembly. Failure of a product originally sold for four or five dollars could result (if there was not any liability problem) in direct charge-backs to the producer of a hundred dollars or more. These specific charge-backs are easy to document; the ill feeling, loss of trust, and deterioration of competitive position from such defect returns are more difficult to quantify but are just as real.

Mechanized assembly can often be fully justified by the reduction in warranty costs alone achieved through total functional inspection of the product during the assembly process.

Reduced Liability Exposure. If the steady drain of warranty costs is a severe drain on profitability, the trauma of product liability offers severe threats to the life of a product or to the manufacturing company. Broad expansion during the 1970's and 1980's of the product liability concept to all involved in "the stream of commerce" mandates a new look at preventive or defensive measures. Liability exposure can result from poor design, poor fabrication, poor assembly, and failure of the product to perform its intended function. Because of legal problems and insurance structures most liability claims focus on manufacturing rather than design defects.

Defense against liability claims must prove that the manufacturer has taken all the necessary steps that a prudent person could take to ensure product quality and function.

Conversion from manual assembly to mechanized assembly offers opportunities for defense against liability claims. The major capital investment required for mechanized assembly justifies an in-depth review of any product design. Not only should the design be analyzed for its ability to be assembled automatically, but ensure that the opportunity is there to analyze the product for any potential liability exposure.

The necessity for consistent component quality required for reliable mechanized assembly will tighten up manufacturing and purchasing techniques.

The uniformity of assembly achieved by mechanized assembly significantly improves overall quality. Studies of product quality enhancement through mechanized assembly of military fuses in one case reduced the requirement for destructive testing of two fuses out of each 100 produced by manual assembly to two fuses out of each 1000 assembled automatically.

The most significant protection of all is through the often inexpensive addition of quality and functional testing stations to mechanized lines and by the methodical logging or recording of the data produced by these stations. This can be as simple as logging counter readouts or as sophisticated as recording data by direct transmission (on demand or real-time) from the machine control system to the management information computer system, even as compliance with ISO 9000 or QS 9000 programs.

Living in an Age of Regulation. The number of regulatory agencies, and their regulations governmental and non-governmental, seems to grow at a logarithmic pace. Non-governmental organizations such as Underwriters, NEMA, ANSI, and EPA are all typical of those entities that stipulate certain industry wide quality standards for manufactured products. While these agencies will monitor performance through selective sampling, they often specify that the manufacturer functionally test all production. Incorporation of testing stations or mechanized lines is not only modest in capital cost, but automatic testing removes much of the subjectivity, inattention, and poor judgment often associated with manual inspection.

Worldwide Competition. Developed western countries and Japan realize that they not only must compete with one another but with the enormous pools of inexpensive labor in the third world. Mechanized assembly offers not only competition to inexpensive labor, but allows a product with verified quality to be manufactured in close proximity to major marketplaces. It often means that manufacturing can be done very profitably in areas with a greater degree of political stability.

Supply Side Economics. One of the fundamental assumptions of supply side economics is that inflation will be controlled and economic stability restored through adequate capital expenditure in productive facilities; that supply can cope with demand.

It is too early to judge whether this theory of economics will become permanent governmental policy. If it does, it will have direct impact on procurement policies for mechanized assembly systems.

Mechanized assembly not only reduces direct and indirect costs but it tends to stabilize assembly costs when other costs are constantly rising.

THE HUMAN FACTOR

In evaluating the potential return on mechanized assembly systems it is too easy to compare projected costs against historical costs. It is not sufficient to include projected inflation in the analysis without considering other human factors that impact on justification for mechanized assembly.

The Declining Rate of Productivity

Many western style countries face a declining rate of increase in productivity, which has its fundamental roots in changing attitudes of industrial workers. The problem is twofold: psychological and educational. Whether it be a lack of motivation, a lack of patriotism, a lack of work ethic, or a different attitude toward the quantity of work done for a specific amount of compensation, workers in the main are not inclined to increase their personal productivity. Those who are inclined to be hard workers are often forced to lower norms by the peer pressure of their less productive fellow workers. (Figure 1-5.)

The productivity problem is compounded by insufficient education, whether formal or on-the-job training. Many modern line supervisors consider their jobs stepping stones rather than terminal positions and often lack the technical skills to educate their charges.

With the exception of Japan, individual firms face an uphill battle in turning this situation around. In fact, decreasing personal productivity is often encouraged by government policies.

With the complete retirement of the children of the Great Depression of the 1930's, vast cultural changes have taken place in industry. Investment in mechanized assembly equipment may be justified as a means of pure survival.

The Age of Consumerism

Another significant factor in looking at mechanized assembly and its capacity to improve and verify quality is that the product consumer is demanding more quality while pride in work seems to be on the decline.

An honest evaluation of the possibility of motivating workers to improve and sustain quality day after day may strongly indicate purchase of mechanized assembly where normal justification for purchase through labor cost reduction alone is marginal at best.

Fig. 1-5. A manual assembly line. Most workers adjust quickly to the norms established by their fellow workers.

Declining Manufacturing Skills

The generation of workers who are currently reaching retirement age were skilled enough to enhance the capabilities of old, obsolescent, and inefficient tools with a minimum of maintenance and by human compensation for the inadequacies of the machinery. This craftsmen's attitude, often handed down in families, has deteriorated to the point that machines must have the control sophistication and physical rigidity to achieve product quality without significant intellectual input from operators.

This replacement of human motivation and skill through modern machines and controls is as essential on the assembly floor as it is in the fabricating departments.

Market Resistance to Increased Prices

In an age grown accustomed to inflation, there is an attitude in many manufacturing firms that encourages passing along manufacturing cost increases to keep profit ratios constant. Each marketplace, however, generally has a point at which further price increases are resisted by the con-

sumer. Major appliances, wiring devices, and, today, automobiles are typical markets faced with severe customer resistance to further price increases. Manufacturers usually try to get additional increases by including highly profitable optional features to the product to obtain market acceptance for higher sales prices.

Where industries face increasing resistance to price increases, cost reduction or cost stabilization becomes critical to survival. The initial cost reduction and subsequent labor cost stabilization achieved by mechanized assembly are significant factors in mechanized assembly system justification.

Modest returns on assembly systems investment may have to be accepted to protect major investments in plant and fabricating equipment.

WHY THE DELAY IN USING AUTOMATIC ASSEMBLY

Despite repeated lip service for almost two decades by governmental agencies, trade associations, and news media to the theme "Productivity," some basic facts underscore a judgment that there remains far more energy going into the discussions of the problems of assembly productivity than into implementing presently available solutions to these problems. Some facts are clear:

a. The rate of increase in productivity in many industrial countries is declining.
b. The largest percentage of remaining direct labor content in manufacturing is in assembly. It exceeds all direct labor costs for fabrication, forming, molding, plating, finishing, and metallurgical treatment combined.
c. The area of assembly and related testing is the only remaining area of manufacturing that is capable of broad scale significant reductions in direct labor cost.
d. The expenditure by industry over the long term on automatic assembly is continuously declining in constant dollars.

The problems of American industry, shared by many of our free world counterparts, are created by continuing social changes unparalleled in history since the Industrial Revolution of the mid-1800's. These problems are compounded by severe changes and increases in governmental controls and continuing obstructions to capital formation. Automatic assembly, successfully applied, can meet these challenges. On the whole, however, manufacturing engineers still find it difficult to convince management through the use of normal accounting practices of the value of automatic assembly, both tangible and intangible.

No manufacturing engineer can expect to be fully successful in selling top management an appropriation request for mechanized assembly systems in today's competitive economy without directly addressing himself or herself to two basic topics and their corollaries.

The first consideration is that the use of mechanized assembly is an optional one on the part of management, and the second consideration is that the investment risk in special machine acquisition is a very real one.

Automatic Assembly Remains an Option in Manufacturing

Purchase of automatic assembly equipment is completely optional on the part of management. No one expects to mold, form, fabricate, heat treat, plate, coat or finish high production components manually. A decision to make the product implicitly means a decision to buy or use the necessary equipment and tooling to manufacture the component or buy the components from a source having the necessary capital equipment. Assembly, however, in almost every case can be done by hand with little or no capital expenditure.

Management is faced with so many mandatory capital expenditures, hostile takeover, mergers, etc., that optional capital expenditures must be actively sold as worthwhile, profitable, and essential. It may seem contradictory to say that an optional expenditure can be essential. Obviously, to make components on a production basis, equipment is necessary. When the equipment is combined into a plant, expenditures for fuel, taxes, and environmental and waste controls are also mandated. These expenditures are necessary to making the product. The ultimate goal is to make a product profitably. In varying degrees, material and labor are optional. Selection of plant sites, labor pools, and material choices are management judgments concerning costs. A greater degree of management options concern advertising and marketing, financial controls, data processing, inventory levels, etc. The management judgment in each of these areas is optional, but may prove essential to ultimate or continuing profitability.

In any corporate entity, the CEO of that entity, whether he or she be titled president, vice president, divisional manager, or whatever, ultimately must make the decisions and accept the responsibility primarily for those options he or she approves and to the degree he or she supports them. Those responsible for marketing, production control, purchasing, accounting, data processing, and manufacturing come to the CEO with their options and plans. Each strives to not only do his or her share, but to excel in contributing to the strength and profitability of that facility or division.

In the nature of things, many American and Pacific Rim CEO's are marketing or financially rather than manufacturing oriented. If they are not, the executive or financial committee of the board of directors is so composed. It is also in the nature of things that those with the responsibility for administrative matters tend to have greater communicative skills than do engineers. It is no wonder, then, that manufacturing engineers often feel that their capital appropriation requests receive scant consideration, while other fiscal requests seem to sail through. Engineers are not often sensitive to the different tax treatment of capital investment.

Appropriation requests for assembly equipment must include selling skills usually not often required for other types of manufacturing equipment. The day is long since past when a short presentation on direct labor reduction is sufficient to guarantee approval for capital expenditure.

Essentially, the justification document should be realistic in terms of procurement and operating costs and conservative in stating probable returns on investment. It should anticipate all possible objections to such procurement, both factual and fallacious, and respond to these objections in a way that is clear, concise, and objective. But most of all, it should emphasize those elements of the proposal familiar to and of importance to those with authority to approve such projects. These include not only contribution to profit, but specific answers to the social and governmental challenges to business continuation and prosperity. Not the least problem is the residual fear of failure following the fiascos of the 1980's in robotic, vision, and guided vehicle investment.

Automatic Assembly Has a Poor Track Record

The track record of many attempts to mechanize assembly has been poor. Dr. Powell Niland published his classic work "Management Problems in The Acquisition of Special Automatic Equipment" in 1961 in which he outlined areas of failure. The book's abstract stated in part:

Because of its distinctive characteristics, the acquisition of special automatic equipment calls for different policies and procedures from those for conventional equipment. This study entailed an examination of the acquisition process in eighteen different plants in order to develop a general description of what was involved in the acquisition process and to identify the locus of major problems.

All of the evidence pointed to the *debugging* of the equipment, after it had been designed and constructed, as the outstanding source of difficulty. The tasks of appraising project proposals, determining the basic methods and equipment specifications, and designing the equipment also developed as major problem areas. The relations between a special automatic equipment group and the rest of a firm's organization were also examined to discover areas in which problems were frequently encountered. Two sets of relations were found to be critical: those with the firm's product design engineering department and those with vendors building special automatic equipment.

The debugging step, making a piece of equipment work efficiently after it has been built, is a fundamental characteristic of special automatic equipment. It therefore became one of the key problem areas singled out for special treatment in this research. Many of the errors and omissions of prior steps in the acquisition process, such as poor design, poor workmanship, and faulty assumptions about parts, processes, and conditions of use, come home to roost in the debugging stage. Even when the preceding steps are skillfully executed, a significant amount of debugging remains because so much of the equipment comprises new or previously untried combinations of mechanisms. To reduce the difficulties in debugging, the author recommends detailed planning and careful execution of the preceding steps in the acquisition process, systematic testing and recording of data, and the use of statistical quality control techniques to identify the causes of defective performance.

One of the most important conclusions emerging is that there are substantial benefits to be realized from careful pre-award planning, especially by working up soundly conceived specifications in considerable detail. Some of the more frequently encountered difficulties, as well as some of the more serious, had their origins in failure to perform adequate pre-award planning. Efficient communication between representatives of the buyer and vendor also emerged as a critical need. Finally, the author concludes that in acquiring this type of equipment there is some advantage in restricting purchases to a relatively few, carefully selected vendors with whom the user can work in close cooperation at all times during the acquisition process.

Still another area selected for special attention was the relationship of special automatic equipment to the product design function. The author's inquiry shows that effective coordination between these two functions is, in the long run, fundamental to a successful special automatic equipment program, and that it should be the kind of coordination which involves mutual interaction and adaptation. This kind of coordination helps increase the number of potential applications for special automatic equipment and it also can help avoid a great deal of the obsolescence costs which are so great a risk. A program of product standardization can help in both these ways. Also, from the standpoint of special automatic equipment design, standardizing the basic machine types and elements using so-called "building blocks" can also reduce the impact of obsolescence charges by permitting the reuse of some portions of equipment made obsolete by product design changes.

This book, now over 34 years old, remains as valid today as the day it was printed in outlining why special equipment often fails to live up to unqualified expectations, hopes which in themselves may be fallacious. The basic question to be answered is, "Is automatic assembly practical?"

Approval cannot and should not be expected for unrealistic programs. Automatic assembly is a systems approach to manufacturing. Additionally, the proposed system must operate within the limits imposed by the manufacturing environment for that system.

Before even going out for quotations on an as-

sembly system, there are areas that must be considered in depth:

a. Do the components of the assembly lend themselves to mechanized handling? If not, can design changes be effected? What are the cost factors involved? Parts configuration, surface considerations, and fragility are among those reasons that may preclude automatic handling of individual parts from bulk condition. If there are such parts involved, is it practical to manufacture, mark, or configure these parts during the assembly process?

b. Is there sufficient volume on a continuing basis to justify the large capital outlay of automatic assembly? A machine that assembles only 10 pieces per minute, produces 1 million assemblies per year. Is required volume affected by seasonal production demands that mandate high production rates for a limited portion of the year?

c. Are model changes and part designs stabilized? Is projected model life long enough for sufficient production after machine installation? If the model life is short, there may be insufficient time to recover invested capital. Most experience indicates that people are unrealistic about the life of a model run. Most products tend to run much longer than expected. With increasing government regulation, it can usually be assumed that approved or certified products will continue in production for rather long periods.

d. Can manufacturing operations and material handling before and after the assembly equipment be coordinated to provide continuous, controlled quality component input and efficiently and safely store the production? An assembly machine tends to be inflexible in its demand for sufficient component parts available at all times. This means disciplined production and inventory control of parts coming to the machine, and efficient material handling at the output side of the equipment.

e. How much direct labor can be saved by such mechanization? How much direct or indirect labor of the higher grade will have to be added for supervision and maintenance? Most assembly builders evaluate requests for quotation on the basis of present or projected labor content. Direct labor reduction will continue to remain a portion of any potential return on investment. Do not neglect possible reduction in various types of indirect labor and support personnel. Reduced numbers of indirect labor such as inspectors, material handlers, and payroll clerks can significantly contribute to the overall return on investment. It must be remembered, however, that such complicated or sophisticated machines will require supervision and maintenance skills superior to that usually required on manual assembly lines.

f. To what extent can in-plant engineering and toolroom capabilities be used in designing, installing, tooling, and debugging automation units? It requires sound judgment to evaluate honestly the existing in-plant capabilities that can be coordinated or utilized in system development, thus reducing the overall cost of the project. The author's experience indicates a common failure to fully utilize in-plant test capabilities. At the same time, we often see a tragic overestimation of capability of designing large assembly systems. Frank discussions with prospective builders can often bring about a happy balance in the utilization of the talents in the builder and user organizations.

g. If component parts to be assembled are expensive, will the relatively high cost of parts used in tooling, debugging, and pilot runs be recoverable? An insufficient supply of component parts means sketchy and incomplete debugging. This puts the burden of debugging on the production facility at enormous burdens of time, frustration, and missed schedules.

Answers to these questions should determine overall feasibility. A brief outline of this examination included as a component of the appropriation request is a valid prelude to reviewing the actual cost bids by proposed vendors. It also is an excellent aid in obtaining the lowest possible bids for good production systems, since readily available concise answers to these questions will indicate to the builder that all aspects were reviewed before issuing the request for quotation.

While it may seem excessive to include a brief description of this feasibility study in a capital appropriation request, it should not be forgotten for a moment that automatic assembly is an alternative method of manufacturing. Management will want a full statement of its options. Technical feasibility must be viewed in the light of overall project and ongoing operating costs as well as both tangible and nontangible benefit.

Determining Project Costs

Once it has been determined that there is sufficient product volume, product stability, technical feasibility, and potential cost reduction, the next step is to determine the overall project cost. There are many cost factors that must be considered *above and beyond the purchase price of the assembly machine*. Consider the following:

a. A projection of the overall customer salary and travel expenses for engineering liaison. Automatic assembly system projects usually require substantial liaison between the builder facility and the user. Station designs must be reviewed, component part vendors' sources may have to be visited, and trips taken to the builder's site during preliminary and final acceptance runs.

b. *A formal review of end item quality control requirements, both present and achievable.* Increasing consumer demand for product reliability should result in a determination to include in the automatic assembly machine every practical quality assurance inspection. Inspection stations should provide not only for the presence and position of each of the component parts, but, as far as possible, for determining the function and reliability of the completed assembly prior to ejection. Inspection station design and data recording of quality as well as trend analysis capability should be coordinated with those specifically responsible for quality assurance at the user site.

c. *A determination of the cost of redesigning existing products to facilitate automated handling.* It may be necessary to redesign individual component parts or total assembly configuration in order to make automatic assembly practical. Take into account necessary die, mold, and tooling changes in order to incorporate the required changes. These may require substantial additional up front costs, but, at the same time, provide excellent opportunities for further cost reduction. For example, a new technology such as ultrasonic welding may remove dermatitis problems, reduce or eliminate cure bonding times, and substantially eliminate work-in-process storage requirements for parts previously bonded with solvents or adhesives.

d. *Identification of the cost of efficient integration with present plant equipment.* Work flow requirements may mandate conveyors, storage elevators, and new pallet or work carrier designs not required for hand assembly. This cost must be balanced against reductions in material handling costs achieved by mechanized assembly.

e. *An estimate of the cost of converting a proposed mechanized assembly system for possible future redesign of component parts or product.* Be sure that quoted machines have salvage value and can be readily retooled or reused in the event the product is redesigned, changed or eliminated.

f. *Personnel requirements.* Supervisory emphasis will shift to technical skills rather than people handling skills. It may be necessary to locate automatic assembly equipment in departments other than the present or potential manual operations.

g. *The various methods of depreciation and available tax credits.* Tax credits are very important since they reflect a direct credit on taxes paid and would have a fairly immediate impact on cash flow. Various depreciation methods can be used to show maximum results depending on corporate investment goals. Much of the cost of assembly equipment lies in the engineering and tooling areas and, hence, as an expensed item, can be quickly written off should this prove useful in a given fiscal year.

h. *The degree of salvage ability in the equipment should the current product be phased out.* Can the quoted machines be used for other similar or emerging products or usefully transferred to other divisions? This particular aspect is becoming increasingly important as major technological changes are imposed by environmental or consumer concerns. (Figure 1-6.)

The cost of these additional expenses must be considered in determining the validity of using a mechanized assembly approach. The total costs may preclude the use of mechanized assembly as a viable option, but they cannot be swept under the rug and must be part of any project consideration.

A prime concern is the capital cost of the assembly system itself. The options in assembly machine design available to a process engineer soliciting proposals are so varied as to require a great deal of analysis. Obviously the less the capital expenditure and the greater the savings, the easier the justification.

Seeking Quotations

In order to ensure the best possible cost quotation, there are several useful guidelines to follow.

Outline the problem rather than the answer. So often assembly system builders receive requests for quotation so specific in details of machine construction and configuration as to preclude any creative input from the builder.

The process engineer faced with choices in automatic assembly equipment might well take the advise of Thomas Edison, "Genius is 1% inspiration and 99% perspiration." Uttered by one of the most creative minds in our technological history, this comment is certainly worth considering. Automatic assembly system development still remains as much art as science.

Experience is an essential factor of successful assembly mechanization. It is necessary for a potential customer to distinguish between a builder's reputation and a builder's record. Reputation consists of impressions and subjective judgments, whereas records are based on factual historical evidence of success or failure. A potential buyer has the right to insist from any builder a list of prior customers and installations. Guarantees as to performance are worth as little or as much as the builder's past performance record and financial strength. Failure to address this experience issue was the basic problem in the automation fiascoes of the early 1980's where internationally known companies entered factory automation without necessary experience or proven successes.

Fig. 1-6. Standard assembly machines under construction. The increased use of standardized machines increases salvage value when a product is phased out, or when significant product changes occur.

It should never be forgotten that once a process engineer or purchasing team commits to a builder, that person has placed his or her own judgment on the line with their company. It is only human nature to attempt to protect one's purchase decision, should the builder fail to live up to the explicit or implicit promises of his proposal.

This may mean taking incompleted or untried machines into the plant, providing unexpected advanced cash payments, and incurring unplanned expenses for toolroom maintenance and engineering time at the user facility.

In examining proposals, a buyer should expect to find a detailed statement of the work to be done by the machine and a specification of the equipment to be supplied to perform each stated function. The cost section of the proposal should include an itemized breakdown of the cost for the various inspection units and inspection functions; for work-holding fixtures; for each feeding, joining, transfer, or fabricating function; and for any specialized ejector or conveyance device, as well as palletizing equipment buffer storage.

Such a station-by-station cost breakdown rather than a lump sum price is some indication that those involved in preparing the quote have been realistic in determining some method of action for each of the described functions. The proposed builder should be prepared to discuss a valid rational design concept for each of the stated functions. This detailed breakdown also indicates the builder can justify on a detailed basis the proposed station costs. *More important, however, should it become necessary to modify any of the quoted functions or stations during the course of the machine building, there is a point of departure for any cost adjustments.*

The builder should also be able to describe approximate floor space, power requirements, payment terms, and gross production rates.

Delivery times are an essential element of any quotation. A buyer has a right to know the builder's current backlog and monthly delivery capability. The buyer should consider lead times on purchased items, and availability of the builder's own standard components necessary for meeting the projected delivery promise. On the builder's side, however, there is also an assumption that prints and/or samples are available to commence engineering at the earliest possible time. This becomes a major problem in any simultaneous engineering project.

The builder should be prepared to discuss in the proposal or in related discussions what elements determine or limit the cyclic rate of the equipment and whether or not any stated manual operator functions can be done ergonomically in a mutually agreeable time span.

In evaluating quotations, the buyer must consider the basic machine chassis or fixture transfer system proposed. Generally, mechanical machines have a greater initial cost, lower maintenance and operating costs, and less downtime. Fluid power machines generally cost less initially, but require higher maintenance and have less up-time.

Control systems and control options are chosen for a variety of reasons. Mechanical integral machines generally need extremely simple controls to index fixtures and to transfer parts and other machine elements, but significant memory or logic circuitry may be suggested for various valid reasons. If the machine requires sequenc-

ing, logic systems are indeed useful tools. If there are a large number of quality control aspects to be considered together with alternative modes of system operation, control circuits using logic functions again may prove invaluable. On the other hand, if memory systems are quoted *solely* as a means of overcoming part quality problems, their cost is usually indefensible in practice, no matter how attractive on paper. This is discussed extensively in Chapter 4.

The buyer's problem, of course, is to ensure that the chosen assembly system has sufficient funding to be built properly, to be debugged thoroughly prior to shipment, and to provide all necessary installation services that may be required to put the machine properly on line. On the other hand, the buyer must beware of unnecessary expenditures and also proposals that appear inexpensive at inception, but continue to grow in size as the project develops. *Over-specification at time of request for quotation is usually one of the main reasons for excessive project costs.*

Customer specifications for machine design are usually the end result of a history of specific problems in the operating plant. Specifications attempt to create a general answer to specific problems. They may unnecessarily complicate good machines or exaggerate purchase costs. A check list type of specification in which the process engineer selects those specifications important to his or her specific project is most useful.

The appropriation request should detail briefly the consideration given to various quoted machine systems and the basis for the choice of the selected system. One of the basic problems in selling assembly systems to management lies in the fear, conscious or subconscious, of possible failure. A concise statement as to the actual record of the selected vendor would be a valuable addition to the appropriation requests. Details of some related or similar successful installations of that vendor can go a long way to easing the fears of non-technical management.

A statement regarding planned provision in the selected machine to handle future model or product changes will aid in relieving the fear that the machine may be obsolete before realizing significant return on investment.

DETERMINING NET PRODUCTION

Up to this point technical feasibility and total project costs have been discussed. There is another important consideration—probable net production.

Few established builders are willing to guarantee specific net production levels on large assembly systems unless such machines are essentially duplicates of proven and tried equipment. There are four controlling elements in determining that portion of gross capability realized as net production:

- The machine as a system.
- The reliability of the individual station functions.
- The motivation of the operating and maintenance personnel.
- The uniformity of component parts supplied to the machine.

For those with a mathematical bent of mind, the following paragraphs express a relatively simple formula for explaining the basis of net production on a specific machine application in an operating environment. Note the largest variable in this formula is the length of the average downtime caused by failure to insert components properly. This downtime factor is controlled primarily by operator motivation and component part quality, rather than machine design, assuming that the machine at time of acceptance had reached satisfactory levels of production and had a strong suitable design compatible with sustained production operations. Note in particular the distinction between efficiency of a machine and net production of a machine.

Expressed mathematically *net production* can be expressed in two steps. It is first necessary to determine machine efficiency in terms of the percentage of machine cycles that will produce acceptable assemblies. When C is the percentage of acceptable parts in each lot of component parts coming to the machine, S is the efficiency level of each work station in performing its own task of selection, transfer, or joining, and M is the efficiency of the basic machine control system in coordinating all of the individual station operations, the probable percentage of machine cycles that will produce acceptable assemblies can be expressed:

$$\text{Machine Efficiency} = (C_1 C_2 C_3 \cdots C_n) \\ \times (S_1 S_2 S_3 \cdots S_n)M$$

It is possible that on a short run basis, or if the quality or efficiency is quite low, any particular machine cycle can be deficient for more than one reason, bringing up the efficiency level. However, on a long run basis the probable efficiency level indicated by the above formula is quite realistic.

Net production across a reasonable length of time is determined in the following way. The number of inefficient machine cycles C_d in a given period (obtained from determining machine efficiency levels as previously described) is multiplied by the average downtime T_d that each malfunction causes. The resultant is subtracted from the total time T_t in the period, and the result is divided by the gross cyclic rate R_g:

$$\text{Net Production} = \frac{T_t - (C_d T_d)}{R_g}$$

We will review these formulas in Chapter 4 since an understanding of the implication of average downtime, random failure patterns, and the laws of probability and possibility must be considered in systems control choice.

Most builders are willing to give probable levels of net production as indicative of attainable goals without giving specific guarantees. An experienced builder will vary his or her estimates based on the assessment of the customer's technical capabilities, quality assurance capabilities, and personnel motivation factors. The builder will also factor into his or her assessment yield levels that will be expected in various industries. Yields generally are lower in electronic assemblies than in mechanical assemblies. Yields will be better in assemblies specifically designed for mechanical assembly than would be true in cost reduction projects on existing assemblies.

Again the machine purchaser must tread a careful path between exaggeration of possible yields from any assembly system and understated net production because of a concern over machine productivity. If such concern does exist, the project probably should be held in abeyance until a probable net production can be better established. Again, a concise statement of the rationale for the projected net production levels will do much to preclude unnecessary resubmissions for appropriation approval.

The Mythology of Capital Expenditure

After presenting a technical and financial justification document strongly indicating the value of purchasing a mechanical assembly system, the manufacturing engineer often runs into frustrating roadblocks, much of which has to do with statements concerning the availability of capital for investment.

While each corporate entity has its own long-range goals for development, very few deviate from the basic investment principle of obtaining maximum return (ROI) on minimum capital investment. Each side of this statement bears on the other. Too small an investment may produce no return whatsoever. Over-investment without subsequent gains in productivity diminishes the potential return on investment. There are practical upper limits on available capital, and return on investment certainly is a vital criterion of investment priority. Too often, however, manufacturing engineers accept without a struggle rejection of their appropriation request based on such statements as the following: "This machine fails to meet our return on investment goal." "Our corporate policy will not permit the purchase of any equipment having a return on investment over X months." "We can only utilize a portion of the indicated savings since we will have to pay income tax on any profits realized from this investment."

These statements assume that the corporation has a wide variety of choices of investments, each offering major returns on investment. Under close scrutiny, however, many of these so-called major returns turn out to be particularly intangible. Many well-run conglomerates are saddled with profitless investments purchased on false hopes. Had this money been spent on improving productivity, their current position might be substantially better. It is much easier for manufacturing engineers to document the reality of potential return on their suggested investments. These are real returns and in an inflationary period tend to grow in value. As an essential part of management, manufacturing engineers have both the right and the duty to question the validity of some of these negative comments.

Management can expect very short periods of return on automated assembly investment only when the alternative existing manual method has been inefficient and overly expensive. Competitively operated plants must expect longer return on investment at each increasing state of mechanization.

Plants that make investment decisions based on present rather than future market conditions tend to ensure that such market conditions will be poor, since they limit their ability to recover from recessions quickly and purchase equipment only in boom periods where deliveries are extended and prices high.

The manufacturing engineer must sell the tools of assembly productivity aggressively and with intelligence and skill. Filling out the figures on simple forms will not suffice under today's present conditions. Appropriation requests must be presented intelligently, attractively, concisely, and clearly.

AREAS OF APPLICATION

Mechanized assembly will not be feasible now or in the foreseeable future for many types of assembly. There are, however, several areas where it should always be considered seriously. These areas include assembly work where many people perform the same operation, where the product to be assembled is relatively small, where foreign competition is intense, or where market dominance is desired.

High Repetitive Labor Content

Whenever mechanized assembly is considered, chances for justification increase where two or more operators are involved in the same assembly task, or when several people are required to obtain the necessary production rate. Justification is extremely rare where one or two people produce all the required assemblies. It is not always possible to eliminate all manual content in assembly, and often justification is best obtained through semi-mechanization and by efficient coupling of people and machinery through operator assist devices and mechanisms.

Small Parts Production

It is in the nature of things that small assemblies are usually more practical for mechanized assembly techniques than larger assemblies; cost of assembly machinery increases rapidly with product size. When parts are small, automatic feeding of components from unoriented bulk condition is usually practical. As component part size increases, costly storage devices, such as overhead conveyors, may be required to transfer components into the assembly line. Such peripheral equipment often raises total project costs to unreasonable levels.

The dramatic increase in cost as assemblies become larger has direct correlation to the size and market share of the producer. In relatively small assemblies, the lower cost of mechanized assembly machinery may be justified readily by very small firms. These firms may make a unique product for a limited market and may be the sole producer of that product. On the other hand, some markets are so large that even 1 or 2% of the overall market may require such high annual production that a small firm with small market share may readily justify assembly mechanization. For instance, a small firm with 1% of the disposable hypodermic syringe market in the United States would need to produce 30–40 million units a year to supply this limited share of the market.

As products get larger (for instance, an automobile), the capital cost for mechanized assembly increases logarithmically. Larger producers can justify equipment much more readily than small producers. The increased efficiency and profitability of major producers increases their ability to dominate the marketplace and thereby increase overall profitability.

Intense Foreign Competition

Global markets and global competition are now commonplace. If foreign imports are seizing an increasing share of the market, mechanized assembly may be one of the few options available for survival. It is imperative that a decision to maintain market share in the face of foreign competition be done while there is still sufficient production to justify proper product assembly systems. Too often the decision to fight is made after production levels have dropped to unrealistic levels for economically feasible mechanized assembly. Market share maintenance is far more feasible than recovery.

Securing Market Dominance

Companies wishing to increase as well as retain market share over the long haul face the inevitable "chicken and egg" question. Should a firm acquire productive capacity and then seek market share, or should market gains be sought and then production capacity built up to match sales. Companies that are run by accountants favor the latter course; entrepreneurs favor the former course. The lead times for developing mechanized assembly systems are such that the choice of increasing market share without productive capacity available usually means a period of 12–24 months of unusually high labor costs, production crises, and delivery problems. Which of the courses to follow will usually be dictated by management style.

SOURCES OF AUTOMATIC EQUIPMENT

Three general approaches can be used in the procurement of automatic assembly machines:

1. Design and build the equipment within the user's plant.
2. Employ an outside engineering or consulting firm to design the machine, with actual fabrication performed in job shops or the company's toolroom.
3. Buy a completed assembly machine from an assembly system builder.

It is obvious that each of these approaches has some salient points in its favor. All have been widely used and will continue to be used to some degree. In judging which is best, the management team should consider the strong and weak points in each approach.

In-Plant Development :
The Do-It-Yourself Approach

The tendency of some few companies to build their own assembly machines can be attributed to several things. Some companies consider this activity a logical development of the in-plant manufacturing-engineering function. Many plants have long dealt with complex transfer dies and press feeders and with the development of special-purpose machines. Also, keen competition within an industry results in a desire to prevent manufacturing procedures from becoming known to competitors. Another major reason is the apparent high initial cost of outside procurement. Outside procurement costs are readily identifiable, while inside building costs are often hidden in a variety of direct and indirect expense accounts.

Each of these factors can be a valid reason for the do-it-yourself approach, if substantiated. However, the following points should always be kept in mind.

First, automatic assembly still is very much an art. Much of the successful inspiration of today is in reality a recollection of the perspiration of yesterday. Management will have to decide whether sufficient engineering know-how is really available within the plant.

Secondly, in comparisons of the cost of outside machine procurement with in-plant construction, realistic overhead charges must be applied to estimated material and labor cost. Availability of specialized equipment and familiarity with vendor sources have to be considered. In evaluating labor-cost estimates, allowance must be made for debugging runs as well as for fabrication.

Thirdly, in-plant design and construction concentrate all the risk associated with any new machine on the user. Two situations would seem to favor in-plant construction. One is a relatively simple assembly where capital expenditure must be kept to an absolute minimum. Here, several commercial items might be procured and the work turned over to a model maker or experimental toolmaker. The other situation occurs where a constant succession of automatic assembly problems or requirements within a corporate organization justifies the establishment of a special machine division. Such a division would

be a complete engineering and fabricating operation capable of meeting all special machine demands of the various plants. It would provide the fabricating facilities and engineering experience of an outside builder but, in theory, eliminate the profit margin paid to an outside group. From a long-range view, however, whether or not such a division offers cost advantages over outside procurement is a moot point. Salary levels and fringe benefits granted in large companies to in-house special machine development may far exceed the profits of independent assembly machine builders.

Using Outside Design Services

As used here, the term "outside design services" refers to firms that offer engineering services for the design of automatic assembly machines but who are not fabricators. This, of course, is not meant to preclude model making facilities for the tryout of design principles. Properly selected, such firms can offer a continuity of engineering experience and a depth of design know-how. Many manufacturers would find it difficult to hire, and impossible to maintain, such an engineering group for an assembly project. On completion of the design, the fabricating job can be turned over to a low-overhead facility such as a local tool and die shop.

Such design services are especially useful where very extensive development work is required or where the ratio of design time to fabricating time is extremely high. Situations of this type occur mainly when production volumes are extremely high or the design of the part is quite stable. Such work may be better suited to design services than to assembly machine builders. Most system builders have established a balance between their engineering staffs and fabricating departments. Altering this balance on the side of engineering is apt to disrupt the builder's operations. Much of the failure of robotic integrators can be attributed to this problem.

Any evaluation of the use of outside design services must cover several points. The range of competency in this field is extremely broad. Because of the small capital investment required to establish such organizations, many competent firms find themselves competing with groups unable or untrained to perform these services. Clear indication of experience and ability to perform should be required from outside design firms. In addition, separation of design and machine fabrication poses many problems. Such work tends to be open-ended on price. Definition of responsibility, unless spelled out completely in the beginning, can lead to time-consuming delays.

It is essential to realize the function and nature of debugging when considering outside design services. Lack of communication between design and fabrication may lead to long delays before necessary modifications are made. Difficult feeder problems must often be worked out in development laboratories. In most cases, this means that transfer station design is dependent on feeder development. Quite often no financial recourse for the customer is available for improper design. Here again, all risk is borne by the user.

Assembly Machine Builders : Systems Integrators

Assembly machine builders are defined as those firms that will design, build, and debug a complete assembly machine or integrated assembly and test system. As a source for automatic assembly machines, these firms offer certain distinct advantages. Primary benefits are fixed price contracts and risk sharing. The great majority of assembly machines under consideration today can be quoted by assembly machine builders on a fixed price basis. This fixed price means that the builder shares with the user the risk associated with machine development. The concentration of responsibility in one source reduces the time required of the user in technical liaison. Since the price is fixed, there is a distinct advantage to the builder in completing the project rapidly.

Because experienced builders have worked on a wide variety of projects, they can often adapt successful past designs to current problems. Frequently, they have standardized and field-tested units for specific problems. They are often able to transfer experience from one industry to another. However, the degree of success in completing and installing assembly machines will be in proportion to the degree of contractual and technical understanding established between the user and builder at the outset of negotiations. The builder has control of only two elements: station design and control design. Quality level of parts coming to the machine and operator ability and attitude are of equal importance. When a builder is selected, the builder must, in effect, become a part of the user's management team.

Assembly machine reception at the operating plant ranges from grudging acceptance to enthusiastic adoption, almost in proportion to the position the user takes toward the builder. As this attitude swings from blunt challenge to full cooperation, prospects for success increase, time for the project decreases, and quoted prices drop.

Management would be naive not to realize that experienced builders are as concerned with evaluating the prospective user as the user is in judging the builder. The builder's prices and estimated delivery dates on identical machines for two different users may be substantially different. This evaluation will be governed by the builder's judgment of many factors:

a. Past experience of the user in automatic assembly.
b. Extent of prior feasibility studies and process information supplied by the user.
c. Degree of understanding the user exhibits for all the factors that govern assembly machine efficiency.
d. Extent of modification possible in parts design or process to facilitate automatic assembly.
e. Stage of development at which the user will take over the debugging responsibility from the builder.
f. General credit and machine acceptance reputation of the prospective user.

Much of the dissatisfaction that has risen from time to time between users and builders has come from honest but erroneous judgments of intent on the part of both sides. There may come a time when familiarity with automatic assembly is so commonplace that implicit understanding may be expected. That time has not yet arrived.

FORCES DRIVING ASSEMBLY AUTOMATION

The American assembly system industry has reached a point of full maturity. It is not only the primary supplier to North American industry but American designed and manufactured assembly systems are in use throughout the world. The mechanisms of automated assembly are constantly enhanced by ever new developments in quality measurement tools and the enlarged capacity and capability of automation control systems. A number of factors that have developed through the recent years are driving companies to proceed directly to automated assembly unless the problems discussed in depth early in this chapter indicate that assembly automation is not technically nor economically feasible. These factors include the ever growing customer demand for improved quality and reliability, the increasing use of computers and sophisticated software in the design of new products, the full utilization of simultaneous engineering practices' and top management emphasis on reducing time-to-market to obtain early leads in market share and retain and enhance those market shares for every type of product.

Increasingly, the emphasis on just-in-time manufacturing has led many manufacturers to depart from the departmental structure based on specific types of fabricating processes, to fully integrated work cell department configuration in which raw material passes rapidly through the fabricating, assembly and packaging processes with a minimum of in-process inventory.

Quality

The global marketplace places first priority on product quality and reliability and often is willing to pay a premium price to manufacturers of products that have a reputation for such quality and reliability. Broad recognition of the ability of automated assembly systems to identify any defect at each incremental step of added value has led to more and more systems being purchased solely on the basis of improved product quality and the anticipated reduction in a variety of warranty costs that such continuous quality produces.

The increasing sophistication of products hav-

ing closely matched components is forcing more and more manufacturers to utilize the technology of automated matched assembly to achieve those operating characteristics that cannot be achieved by manufacturing fabricating tolerances alone.

Design for Manufacturing

The entire area of designing for manufacturing originated from two fundamental sources. The first was that of early attempts to automatically feed components into assembly systems. It soon became apparent that the reliability of automatic feeding was extremely dependent on availability of components whose physical shape and center of gravity were designed not only for the function of the product but to also facilitate mechanized feeding.

The second driving force to utilize design for manufacturing came about from management recognition that assembly was the point at which defects in fabrication could be separated and where product reliability was determined.

The rapid emergence of analytical software for computer aided design activities has greatly simplified, when combined with assembly simulation techniques, designing products directly for mechanized assembly without any period of manual assembly. This capability often drives the decision to automate without any real consideration for manual assembly.

Simultaneous Engineering

Most manufacturing companies today practice some form or degree of simultaneous or concurrent engineering. The utilization of this methodology very often brings assembly system integrators into the simultaneous engineering team at the very initial concept stage. This generally means that assembly automation is dictated from the very beginning of the project.

Time-to-Market

A recent chief executive officer of the Ford Motor Company, Red Poling, has been quoted, "Whereas the battleground of the '80's was quality, the competitive battleground of the '90's will be product development." The forces that drive time-to-market activities are a recognition of the profitability resulting from achieving and maintaining significant market share. This issue, described earlier in this chapter, stimulates the turn to assembly automation as more and more companies recognize that effective time-to-market activities also require a significant manufacturing capability to respond to market share opportunities.

Work Cells / Integrated Manufacturing

Concurrent with robotic development in the 1980's were those of automated material handling systems such as automated guided vehicles and computer controlled warehousing. This tendency had a relatively short life for a significant number of reasons. A growing recognition that many quality problems occurred during material handling operations between fabrication and assembly as well as a recognition of the costs involved in any significant storage of component parts has led many companies to physically restructure their factories into work cells combining fabricating equipment, computer controlled buffer storage, assembly, and packaging in a single department. The benefits of such work cell configuration, where secondary operations do not preclude such layouts, is usually done only when the assembly function is automated. Such work cell layouts dramatically reduce the damage done in handling operations between fabrication assembly and packaging.

SUMMARY

Mechanized assembly is a multifaceted tool for management to increase profit by improving operating efficiency. The nature and degree of mechanization will change with the viewpoint from which automatic assembly is viewed. It can be a simple labor cost reduction project. It can be a major revision of management attitudes toward manufacturing, marketing, and inventory control. Full implementation of successful mechanized assembly will inevitably, and not without strain, increase the efficiency of every area of operation. How this will be achieved is outlined in subsequent chapters.

REFERENCES

1. R. D. Buzzell, B. T. Gale, and C. M. Sultan, *Harvard Business Review* 53 (1), 97–105 (1975).
2. Felix Giordano, "Will It Make Money?" *Manufacturing Engineering* (April, 1966).
3. P. Niland, *Management Problems In The Acquisition Of Special Automatic Machinery,* Harvard Business School, Division of Research, Boston, 1961.

Chapter 2
Product Design for Automatic Assembly

INTRODUCTION

Prior to the 1980's product design decisions were historically based on three major factors: product function, product life, and component cost. Assembly labor was considered a minor problem. Availability, cost, and motivation of assembly labor changed this outlook. Most importantly, consumer demands for quality have radically changed the design decision process. Efforts to mechanize the assembly of products, however, were often thwarted by the fact that in the past little effort was given in product design to the problems of product assembly.

The dexterity, flexibility and ingenuity, or in the present jargon, agility of the human assembler are difficult to duplicate in any automatic equipment. To replace the human hand and brain with mechanical, pneumatic, and electronic devices has required significant changes in component and product design for mechanized assembly to be both technically and economically feasible. Broader use of mechanized assembly has also required a greater use of component and product standardization with differences often restricted to material and cosmetic considerations.

There are no universally applicable rules of thumb to simplify all product design require-

ments for automatic assembly. What is needed beyond a best practices approach is a sympathy for the limitations of assembly equipment to feed, sort, and orient components automatically at high rates of speed. It is also important to recognize that failure to design *components* properly for reliable feeding may be overcome by a liberal application of money. More expensive or ingenious feeders, escapements, and transfer and orienting devices can often overcome component design faults.

Poor product design of the assembly itself, however, may not be susceptible to the same fiscal cure and may never be capable of reliable mechanized assembly no matter how generous the investment. Overall product design must also be interactively involved with quality and parts procurement practices. These, in turn, will determine automatic assembly system output.

This chapter examines all of these areas. The solutions offered are not laboratory solutions, but practical ones. They accept the fact that *product function and market acceptance are paramount*. They recognize that component characteristics and tolerances, particularly with multiple vendor sources, will vary. The suggested solutions attempt to keep component costs as low as possible, but, at the same time, look at component costs in light of subsequent assembly costs. Finally, this chapter also examines product design for its adaptability to quality monitoring during the assembly process. Total unit cost must include wasted value added, salvage and warranty expenses as well as original manufacturing costs. Product design decisions must address cost accounting intangibles such as customer perceptions of corporate attitudes toward customer satisfaction.

PRODUCT DESIGN FOR MECHANIZED ASSEMBLY

Until the early 1980's, product design efforts to facilitate mechanized assembly were concentrated on modifying or redesigning components of the assembly to facilitate automatic feeding. This is truly desirable, but it is only one of several basic considerations if a product is to be designed for automatic assembly.

A more fundamental concern is the nature of overall product design and its impact on technical feasibility for mechanized assembly, systems selection, and available control options. Bad component design can often be overcome by spending more money on feeders, escapements, and transfer devices or by fabricating that component as part of the assembly process. No amount of money will compensate for overall product design shortfalls that preclude consistent, reliable or profitable mechanized assembly. In the following discussions, attention is given to a number of facets of overall product design as they relate to the assembly process.

Types of Assemblies

Each assembled product usually falls into one of two basic types of assemblies: additive or multiple insertion. Some small percentage of assemblies combine both types or requirements.

Additive Assembly. An additive assembly is one in which a series of discrete parts are added to one another in a specific sequence. Failure to insert any component or failure to place components in proper sequence and in proper physical relationship to other components means that the product is defective immediately, and little if any

benefit can be expected by attempting to continue the assembly process once the failure to insert any given component is detected. Typical examples of common additive assemblies would include automotive thermostats, tape cassettes, cigarette lighters, and retractable ball-point pens. (Figure 2-1.)

This type of assembly may or may not include the ultimate housing, cover, or base of the assembly product. Often, good product design will allow a series of subassemblies that are later installed into a final assembly.

It is in the nature of such additive assemblies that some assembly system control decision must be made once the assembly equipment detects any failure to properly insert a component. How this may be done will be discussed at full length in the chapter on control systems. At this point, however, it is sufficient to recognize that *in additive assemblies, failure to insert any component in proper position relative to other components means the assembly sequence should be stopped* until some form of corrective action has taken place. It is in the nature of additive assembly that no further assembly is attempted till the defect is corrected.

There are three corrective actions that can be taken upon detection of this misplaced or miss-

Fig. 2-1. An additive assembly. Failure to insert any component makes it senseless to add other components.

ing part piece. The missing part may be replaced on the machine by manual techniques; it can be handled by continuing to move the defective, incomplete assembly through subsequent stations on the assembly systems, while stations downstream of the failure lock out automatically so no value is added to the incomplete assembly. Segregation at ejection of incomplete assemblies, of course, is necessary if this choice is made. Once, more complex control technology was required for this option; today's control systems handle this with ease. The benefit of such an approach is, at least theoretically, potentially increased production. A third simple approach is to have the machine attendant immediately remove the incomplete assembly at the point the incomplete condition is detected and manually replace it with a good subassembly complete to that point so that the machine can be restored to full automatic operation.

There are three design corollaries in additive product design.

1. Assemblies should be designed to facilitate the monitoring of the assembly process as it occurs in each incremental step of the machine sequence.
2. Access should be provided in the design for sensors to determine presence and correct relative position of each component immediately after each insertion or joining operation.
3. Reference locations or surfaces should be included in the assembly design whenever inserted components have less thickness or size than possible height stack up of dimensional tolerances in parts previously assembled.

Assemblies should be so designed that easy manual or automatic removal of the incomplete assembly is practical. One often sees assemblies that can be easily ejected from the assembly fixture only when the assembly is fully complete.

Wherever practical, product design practice should eliminate the necessity to insert fasteners and join in the same station. Repair and salvage potential is often enhanced if insertion and joining operations are isolated. Only in that way is it practical to detect relative parts' presence before attempting joining.

Multiple Insertion Assembly. A multiple insertion assembly is one in which a series of discrete parts not touching one another are assembled in different locations on a common base, but the success or failure of any insertion or joining operation will have no direct effect on other subsequent inserting operations. Typical multiple base insertions would include circuit boards with axial lead or surface mount components, multi-prong electrical connectors, vacuum tube sockets, and similar objects. (Figure 2-2.)

The machine design implications of multiple insertion type products are enormous. If each of the components inserted on the product base is different in shape and configuration, the basic assembly system need not be much different than that designed for additive assemblies.

Where a common or identical component is repeatedly inserted in a multitude of locations, annual volume requirements will have much to do with machine selection. If the required product volume is low, a positioning table indexing in two or more axis under a single station inserting head is most practical. When production requirements are higher, however, it may be necessary to do a series of inserting operations concurrently, often by means of redundant (or even multiple) tooling. Tooling costs increase rapidly.

Multiple insertion products should be designed so that off line salvage of assemblies with one or more missing components or with a defective joining operation can be done by hand or by simple single station equipment. Since the failure of any given insertion operation does not affect the quality of subsequent operations, simple inexpensive salvage eliminates the need to lock out downstream operations after an indication of failure at a preceding station. Simple provision for segregation for salvage rework may be all that is required.

When high production volumes indicate concurrent insertion at multiple redundant stations, much thought should be given to keeping pitch or parts location large enough to provide for inserting tools, staking units, welding electrodes, and other joining tools of sufficient size and strength to be production-worthy.

Many assemblies will, of course, combine features of both additive and multiple insertion as-

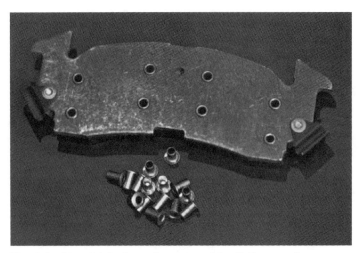

Fig. 2-2. A multiple insertion assembly. Failure to insert any given rivet has no effect on other rivet insertion operations. Salvage by adding a missing rivet is simple and inexpensive.

semblies. In almost all cases, however, one or the other will be dominant, and this dominance together with annual volume requirements will have much to do with the final assembly system choice.

Features for Efficient Mechanized Assembly

In evaluating the practicality or technical feasibility of any assembly project, the experienced assembly system builder will look for several aspects of product design. These aspects should be of major concern and provide guidance to product designers. Most elementary is the inclusion of features that permit the part to be fixtured adequately for automatic operations.

In many cases, functional requirements, the way in which the product is utilized, and aesthetic considerations *may not require any great correlation of external features to internal features.* Take several examples. An internal combustion carburetor needed no close correlation of the fuel intake location to the holes used to mount the carburetor on the intake manifold, or head casting. Its mounting holes have no close functional relationship to its intake jets, choke valve, or other operating features. It is these external features,

however, that are used to locate the carburetor during automated assembly which in turn requires close locational tolerances on internal features, purely for assembly efficiency, not for product function.

Another example might be the exterior door handles once common to all automobiles. The shiny chrome exterior had little functional requirement to be closely dimensioned to the mounting holes, but during automatic assembly the release mechanism located in the door handle (the push button assembly) had to be placed in a cavity that could only be located by fixturing on highly configured, irregular outside surfaces of the door handle casting or forging.

Product designers must incorporate dimensional tolerances or locating features on exterior surfaces of the fixtured component to the degree that internal cavities can be found accurately, and that inserting, joining forces, or pressures can be properly absorbed. In addition to this fundamental design problem, there remain several other areas that are critical to automatic assembly. These include planes of access for component insertion, modular subassembly design, increase of subassembly integrity during the assembly process, and possible use of nonfunctionable, reusable shop assembly aids.

Planes of Access. When parts are assembled by hand, manipulation of the assembly so that parts can be inserted into the product housing or base from any plane of access is inexpensive and, for most products, completely practical. Once automatic assembly seems practical, product designs should be reconsidered so that they limit the insertion of component parts to as few planes or directions of insertion as possible. (Figure 2-3.)

It has been reported that almost three-quarters of manufacturing costs involve some degree of material handling, and few would argue with this assessment. Material handling within a mechanized assembly system is also costly. Most assembly fixtures provide ready access from the overhead vertical position, one horizontal face, and, in certain types of machines, some degree of access for insertion and joining activities from the inverted position. The remaining work fixture faces, those in the direction of fixture travel and interior to the assembly machine, are usually inaccessible. Complex assembly work on the lower face or horizontal faces of any assembly usually involves manipulation of the assembly during assembly and replacing it in separate nests so other faces may be exposed for additional insertion or joining operations.

If component parts must be inserted in several planes, extra transfer stations will be required, additional inspection stations must be added, and extra fixture nests must be included to receive the

assembly in different attitudes. These additional fixture nests add weight to the fixture pallets or carriers and additional inertial loads to indexing machines, while they are only used for a relatively short period of time during transit of the developing assembly through the machine system.

There is a further cost implication when it becomes necessary to revolve, rotate, or reposition a partially assembled product to gain access for additional insertion operations. The design of the assembly at that point must provide for internal retention of previously fed components against gravitational and inertial forces when the assembly is moved to a new attitude. In designing to facilitate mechanized assembly, component parts should be inserted ideally from one direction. If this is impractical, the planes of access required for component insert should be limited to the fewest possible directions.

Modular Design. The product designer attempting to design products for mechanized assembly must recognize that the algebraic laws of probability and possibility have a direct impact on assembly machine productivity, particularly in matched assemblies. This topic is discussed in Chapter 4 in far greater detail. For the product designer the immediate implication is to design complex multi-component products so that they consist of subassemblies that usually should have less than 12 or 13 parts each which then can be

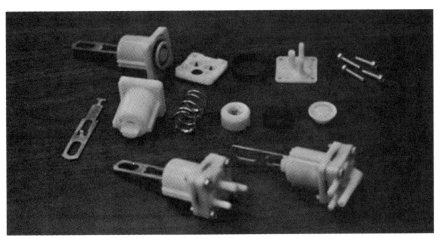

Fig. 2-3. This product is well designed for mechanized assembly in that all parts are inserted from a single direction.

combined into a final assembly. These smaller subassemblies will reduce the number of parts being assembled in any given assembly machine to reasonable limits and ensure (with proper banking or queuing) that the overall line production rate is maintainable. Attempts to assemble more than a dozen or so different mechanical parts on a single machine usually results in significant downtime on the assembly machine. (Figure 2-4.) Electronic assembly systems routinely handle significantly higher numbers of components.

In recent years, so-called clam-shell product designs have enjoyed great favor in order to facilitate manual assembly. Appliances and powered hand tools are typical examples. Such designs applied to complex products (and particularly where internal wiring is facilitated by such design) *are not well suited to mechanical assembly systems,* especially when yearly production requirements are high or where total inserted component parts might number 20, 30, or more. (Figure 2-4.)

Subassembly Integrity. A corollary of modular subassembly design to improve assembly system

net production rates is that *subassemblies should develop sufficient physical integrity from their design to allow automatic handling between the subassembly machine and the final assembly machine.* Theoretically, at least, finished subassemblies should be ejected in an oriented condition and that specific orientation retained in conveyors leading to the final assembly line. In practice, however, such direct linkage may not be satisfactory for many reasons. It will be most helpful if completed subassemblies have enough physical integrity as a subassembly to permit them to be ejected to bulk condition and reoriented in part feeders without physical or functional damage.

Use of Shop Aids. Sometimes, functional requirements and other design constrictions of a specific product do not lend themselves to modular subassembly, since nothing in the function requirements of the subassembly tends to hold the subassembly together. If so, the designer may have to consider the possible value of introducing nonfunctional shop aids such as clips, roll pins, etc. Such shop aids while nonfunctional are inexpensive, and possibly reusable, components that are introduced into the assembly sequence in order to produce a strong subassembly which can be safely stored and transported between the subassembly and final assembly sequence. Obviously, easy removal of the shop aid is usually essential at the beginning of the next operation. (Figure 2-5.)

Families of Assemblies. Relatively few products are made in such volume that dedicated assembly machines are economically feasible for a single product model. During the robotic phase of the early 1980's one often read of the differences between "hard" and "soft" automation. In every type of product, from automotive components to consumer disposables, most production is done in batch quantities. In the past, product designers enjoyed the luxury of designing each product model to suit the marketing department opinion regarding functionality, appearance, and market sector. Each product model was designed in relative independence of other existing or projected models. Few companies insisted on stan-

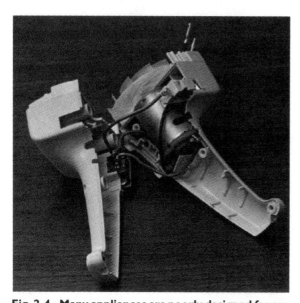

Fig. 2-4. Many appliances are poorly designed for automatic assembly because of the number of parts and the lack of physical stability of many components.

Fig. 2-5. The plastic plug is used to protect fragile components and to provide physical integrity to this brush holder assembly while it is transferred to another assembly machine. This nonfunctional shop aid is used over and over again.

dardization of components and subassemblies used on various models of their complete product line. The appliance industry was typical of many industries in this regard. Various products have been offered in promotional, economy, standard, and deluxe varieties, each having completely different components.

Economic realities have long caused companies to reexamine product lines so that whole families of products can be assembled on a common machine system with little changeover time required. In designing products where there are a family of similar or related models, common locating points and common components and the ability to add or omit specific components can make flexible assembly an economic feasibility where none existed before.

COMPONENT DESIGN FOR AUTOMATIC ASSEMBLY

Most academic and industrial research and study in the past has centered on redesigning

components so that they might be more reliably fed and transferred. Some proposed solutions lie in the area of component shape and weight reduction and the presence of external features. These traits are often lumped into the term "hopperability." Other considerations involve mandating nonfunctional manufacturing tolerances that definitely affect hopperability. Some component parts cannot be fed, at least in the present state of the art, from the discrete state and must, therefore, be partially fabricated and retained in strip form to the point of insertion. In this event, product designers may become more involved in the selection of fabrication processes to a degree not required in manual assembly. The solution to a difficult feeding problem may not require a design change but may involve the way a component is made and the stage of completion at which it is transferred into the assembly under construction.

Hopperability

Most parts used in small product assembly lines traditionally have come to the machine in bulk lots with random orientation of any individual component parts. While all component parts had a specific orientation at time of fabrication, this original attitude was lost at time of ejection or in subsequent deburring, plating, or secondary operations and in storage and transportation from fabrication to assembly. It remains very common to find assembly being done at sites remote from the fabrication. Additionally many assembly components are purchased in bulk lots from specialized outside vendors.

A significant identifiable cost of any mechanized assembly system lies in the development of reliable component part feeding systems. Such parts feeders (mechanical, vibratory or centrifugal) select component parts using characteristics of the components themselves to assist in capturing and feeding parts with specific and eventually usable insertion attitudes.

This inherent ability of any product component to be fed reliably and in specific attitudes is often referred to as "hopperability." There are many facets to hopperability. One essential element is the ability of any parts piece to be fed in a consistent attitude. It is wrong to assume that part feeders orient parts. It is more correct to say that feeders present randomly oriented parts to selection devices that allow specifically oriented parts to pass through and return others of different orientation to the bulk storage area of the feeder. Once initial selection of orientation is made, some degree of reorientation may occur within the parts feeder.

These selection devices are usually mounted in the part feeder in ways that increasingly qualify the random unoriented attitudes of parts as they leave the bulk storage area of the feeder until the selection devices can capture sufficient parts in a usable orientation. This is discussed in detail in Chapter 5. The product designer need not be a specialist in feeder development, but he or she should have some understanding of parts behavior in the feeder bowl. (Figure 2-6.)

Symmetry and Asymmetry. As a general rule, a product designer should strive in component design to achieve complete symmetry or significant asymmetry in each component. It must be emphasized repeatedly, however, that there are three areas of concern in automatic parts feeding: the ability to capture parts in a uniform attitude (not necessarily the desired insertion attitude), the ability to feed parts at a consistent rate in excess of final machine or system cyclic rates, and the ability to feed parts reliably without jamming in selection gates, feeder retention devices and in the final discharge track.

Symmetry alone will not guarantee any of the three. The most symmetrical part imaginable is a ball or sphere. Balls with a minimal surface contact point do not lend themselves well to simple vibratory feeders and often wedge in gravity feeders. Cylinders with a length-to-diameter ratio of one-to-one are particularly difficult to feed and select.

Parts with slight asymmetry or with external symmetry, but internal asymmetry, offer real

Fig. 2-6. Parts traveling in a feeder bowl are oriented by protrusions and other external features touching selection sections of the hopper track.

challenges. The location of the center of gravity in such a part often has much to do with the ease of feeding.

Radial location of internal features in an externally symmetrical part may require custom designed orienting units. Additional station room often can be obviated if functionally critical internal features can be duplicated in the part so as to make it symmetrical. This must be balanced against secondary costs of such redundancy.

Orientation Features. Feeder manufacturers look for "handles" on a component's external orientation surfaces—cavities, or protrusions that will strike or engage selection gates and facilitate part orientation. The flange or head on a binder or fillister head screw is a functionally useful feature. The slot in a slotted set screw is also a functional feature. What is foreign to most product design departments is the introduction of such features that are nonfunctional, but aid automatic part feeding. These really are no different than construction holes or machining pads on exterior surfaces of castings so often found to facilitate parts fixturing and fabrication. (Figure 2-7.)

External orientation features are critical in the case mentioned earlier, where a symmetrical or uniform rectangular exterior conceals an internal asymmetry. Design efforts to facilitate automatic feeding and assembly can either be along the lines of making the interior symmetrical or in providing a nonfunctional exterior selection feature, to facilitate orientation, thus reducing fixture and station costs.

Parting Lines, Sprues, and Gating. Many new product assembly projects utilizing concurrent or simultaneous engineering require assembly machine development and part feeder development concurrent with part fabrication tooling development. The assembly machine builder often must begin his or her work with part drawings of stampings, castings, and moldings and often must begin hopper development with parts produced on temporary single cavity dies and molds. When production parts arrive, frustration may result as parting line position, sprue break-off points, and mold identification numbers are found to be located in areas of production parts which have been previously selected for staging or on surfaces or features used for selection or tracking in the part feeders. Product designers (and mold designers) must begin to consider the impact on hopperability of the location of parting lines, gates, and mold marks of molded components. (Figure 2-8.)

Part Retention by Strip and Web

No theoretical discussion of assembly goes very far without the discussion turning to the possibility of directly coupling the fabrication of components to the assembly line, thus obviating the necessity to reorient parts. This theoretical situation becomes most practical, however, when dealing with parts that cannot be successfully and/or economically fed from bulk condition for many reasons. Such parts would include very thin parts, flexible parts, and parts that would interlock or nest, thus precluding or limiting the use of automatic part feeders. (Figure 2-9.) Such parts are best fabricated and carried to the point of transfer in the remaining strip or web of the original coil stock. Simple parts such as gaskets, dust or static shields, and felt parts can usually be blanked out from stock directly on the assembly machine. (Figure 2-10.)

Those products involving wire terminals have long utilized terminals carried to the assembly line in a preformed condition, but still attached to the carrier web. These terminals are cut off

Fig. 2-7. The slots in the skirt of this piston were added to allow this part to be radially oriented so that the position of the rectangular cavity was established in the vibratory feeder.

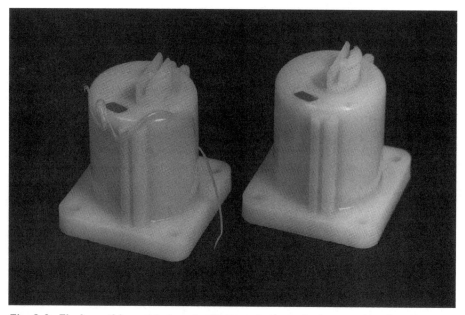

Fig. 2-8. Flash on this molded part will cause jams and stoppages in the vibratory feeder.

and formed to the wire in one operation. One such example may be miniature electronic relays or such parts often referred to generically as "AMP" terminals. This technique has a much broader area of application however than the above examples. The basic questions involving the product design regard the complexity of the die work involved. Simple one or two station dies may permit the part to be punched directly on the assembly machine from strip and carried directly into the assembly by the punch.

If more complex forming or compound die work is involved, a judgment must be made whether to directly couple the press to the machine or to rewind partially formed parts carried in strip or web on reels from the primary press operation for later transfer to the assembly line. The use of reels of partially formed components isolates die maintenance, stock reel changeovers, and other work stoppages from the assembly line operations. Since most press operations run faster than assembly operations, isolation of press work from assembly through the use of transfer reels is extremely practical.

There are several problems, however, with the use of transfer reels of partially completed components particularly if the parts carried in the strip tend to interlock when wound on the carrier reels. Paper interleaving is one solution. A more imaginative solution is to leave the final forming operations of configured components to the point of cutoff and transfer, using the primary press operations solely to blank, draw, and perform. This decision is a critical one since many forming operations that are simple enough when the piece part remains in the web become much more difficult when the part is discrete.

Sometimes final forming is left until the component is inserted or joined into its base or housing. Access for support or forming anvils must be provided in the design if the base or housing has insufficient columnar or compression strength to provide an anvil surface for forming. Unfortunately, such base material might be porcelain or thermosetting material, which can be quite brittle under compression loads.

What is really important is that design for automatic assembly may require a change in the fabrication process rather than a change in the final component shape or design.

Fig. 2-9. Retaining rings are typical parts that cause great difficulty in feeding from bulk condition. Preoriented parts can be fed by magazine units.

Problem Areas

There are many problems in automatic part handling that provide stumbling blocks to machine development. One involves the asymmetrical marking of product numbers or logotypes on an otherwise symmetrical part. Since it is very costly and often impossible to orient such parts reliably, introduction of a small physical asymmetry may be required for orientation purposes.

When considering fabrication on the assembly machine, grain directional requirement and commercial availability of required strip stock sizes may have significant bearing on station layout and machine cost.

In all these areas, the cost of redesigning for assembly can only be judged in the light of overall costs, including operating efficiency and realized net production.

DESIGNING FOR AUTOMATIC JOINING

Ideally, products should be so designed as to eliminate joining and fastening operations by uti-

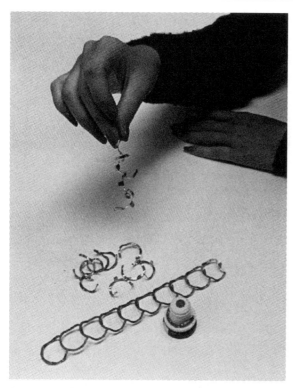

Fig. 2-10. These contact springs for fuses are best handled in strip form after preblanking and forming, with final cutoff on the assembly machine.

design that permits the fasteners to be fed into cavities and inspected before joining operations occur can do much to reduce scrap rates and improve overall efficiency. (Figure 2-11.)

One area that causes problems repeatedly is the tendency of many designers to insert tubular or semitubular rivets into assemblies in a direction opposite to the main direction of component insertion. Other riveted designs require the utilization of riveting mandrels as fixture pins. This practice may suffice for single-station riveting machines, but is not a good approach to multiple fixture assembly machines. One broken fixture mandrel can shut down an entire system.

Adhesives and epoxies requiring pressure application during long curing times create additional costs (and even prohibitive costs) to assembly machine programs.

Each type of joining presents different problems best discussed in Chapter 7. Joining is probably the area where the greatest amount of unsalvageable scrap is generated. Brilliant work in component design can be totally vitiated by failure to consider design options in joining in the light of automatic system assembly requirements.

lizing snap fits or interference fits of component parts. Joining operations and test operations, unlike insertion operations, usually require specific time periods and often may ultimately determine machine final cyclic rates. The types of joining and fastening systems that can be best utilized in automatic assembly are discussed in Chapter 7.

Product designers should keep in mind several considerations unique to automatic assembly. As far as practical, good assembly system design tries to isolate each individual incremental operation of the assembly process. Every attempt should be made to isolate part transfer and part joining operations into separate stations. This preferential operational sequence means that a product should be so designed that the presence and position of each component may be inspected before any joining operation commences. This separation of part transfer from joining operations can be extended to all types of mechanical fasteners, such as screws and rivets. Product

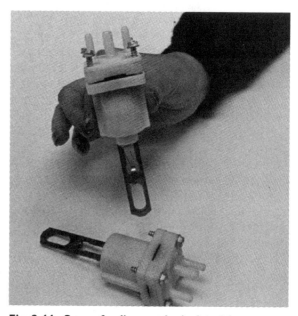

Fig. 2-11. Screw feeding can be isolated from screw driving when product design provides cavities for the screws.

QUALITY REQUIREMENTS FOR AUTOMATIC ASSEMBLY

Quality control or assurance usually means the verification of product conformance to design specifications. Such specifications have been historically based on functionality of the product as dictated by the marketplace, although governmental, legal and social pressures play an increasingly significant role. Automatic assembly introduces a new factor because the first consumer of fabricated product components is the assembly line itself. Product design intended to assist in the mechanized assembly process must review this problem in detail.

Measurement of Part Quality in Assembly Terms

In the intervening years since the publication of Powell Niland's book, mentioned earlier in Chapter 1, and, more specifically, IBM Engineer Arthur LaRue's milestone technical paper, which pointed out the direct correlation of component part quality (consistency) to assembly machine efficiency, there has been a halting but steadily growing recognition of this problem. However, this recognition of problem existence is often without a full understanding of the nature of the problem itself. The increasing use of matched components to meet performance specifications only increases this problem.

For a better comprehension of this problem, it is necessary to briefly review three topics:

1. An understanding of part quality definitions and their intent.
2. The implications of a sigma curve distribution of part variations and the imposition of controls on this curve.
3. The impact of algebraic laws of probability and possibility on assembly machine operations.

We have a habit of using theological terminology for quality control purposes: Parts are "good" or "bad." When we refine our usage, we should say that parts are functional or nonfunctional; within or without print specifications; or within or without design parameters. Parts within design specifications are supposedly considered acceptable (for use), while the others are rejected (unless production schedules prevail).

The characteristics we measure to determine if parts are usable include dimensions, density, hardness, viscosity, electrical resistance, impedance, capacitance, temperature, flexibility, ability to withstand fluid or gaseous pressure, or resistance to compressive, torsional, or tensile forces. These characteristics are measurable in commonly accepted quantitative terms such as ounces, inches, ohms, and the like. Other specifications slide away from the objective to the subjective. Color, appearance, cleanliness may range from the objective to personal aesthetic judgments. Design specifications in themselves may lack specific quantitative value. It is not uncommon to see "part must rotate freely" or assembly "must not leak" or "parts must be free from blemishes or defects" on component and assembly drawings even to this day.

The purpose of specifications for component parts of any assembly is supposedly to ensure functionality of the assembled product within an acceptable cost framework. In general, part specifications should be as broad as possible, keeping in mind the capabilities of part production equipment and functional requirements dictated by the intended use, life-span, and probable operating environments of the assembled products.

Unfortunately, in real life, detail drawings and definition of component and assembly specifications are often left to junior engineering personnel. Dimensions and specifications are often lifted from previous drawings, similar parts or experimental models. The dimensions or specified characteristics may have little or no bearing on available fabricating equipment limitations or actual market requirements or expectations. We come to an important consideration when discussing part quality as it relates to assembly:

Under the pressure of production schedules and fabrication limitations, actual production parts may vary widely from part prints or written specifications particularly if these specifications do not reflect actual production capabilities or functional requirements of the product.

In evaluating assembly machine feasibility, the system integrator must provide for the actual or probable characteristics of component parts, not merely the written specifications.

Determination of probable parts characteristics is a serious problem if assembly equipment is being built in a concurrent engineering environment for a new product concurrently with production tooling or product development. As a rule of thumb, most industries informally establish certain quality levels among the various competitors. For one instance, circuit breaker manufacturers must meet Underwriters' Laboratories specifications and at the same time meet competitive pricing in the marketplace if they are to survive. These commonly accepted industry practices for given commodities dictate certain product quality levels within specific product lines.

It is also increasingly true, in fact almost the norm, that many engineers and tool designers move from job to job but also tend to move within certain industries from one competitor to another, bringing a degree of standardization within manufacturing. Those engineers who prefer or can find occupational stability within one company in turn tend to review previous experiences in tooling up new product components, and this also leads to quality uniformity. In addition, most industries tend to buy production machinery from common sources, and this in turn leads to product uniformity. For all these reasons, if no production parts are available for evaluation of an assembly project, particularly concurrent engineering projects, the safest way to proceed is to determine quality levels in previous similar products by the manufacturer and in similar competitive products currently in the marketplace.

As quality control groups have upgraded their own profession, they have turned their orientation from mere inspection to statistical determination of quality control and then to today's emphasis on quality assurance. This has been an evolution from the identification of defective parts to forecasting of the probable level of defects to an attempt to ensure the absence of defects through prevention rather than detection. This evolution has been a switch of emphasis from meteorological to statistical to managerial tools. Each company is somewhere along this evolutionary path.

In the past many types of statistical sampling techniques were developed. Early methods were designed to identify machine capabilities in the production of component parts. Development of the AQL (acceptable quality level) and similar sampling methods switched the emphasis to product function. Defects were identified as minor, major, or critical as it was determined what specific deviations from individual specifications would have on product function and product lifespan.

It is interesting to note that at one point potential assembly machine users who then become aware of the correlation of part quality to machine productivity often requested specific net production guarantees based on sample component availability of parts with known AQLs. Herein lies the fallacy:

The reliability of any assembly system is dependent on the ability of the machine to feed, select, orient, and insert component parts in a specific sequence to a specific spatial relationship *often utilizing dimensional features for orientation, transfer and staging during the assembly process that have no necessary correlation with product function.*

There are a number of corollaries that can be drawn from this fact. Among them:

- Parts passing all quality control criteria for product functions alone may not have the consistent features necessary for automatic part orientation, selection, and transfer.
- Parts that will not make a single functionally good assembly can often be reliably assembled from a pure physical standpoint, particularly if the functional defects are nondimensional.
- Component parts meeting all specified quality control criteria may not necessarily produce a functional assembly.
- Functional assemblies can result from the proper mix of unacceptable components.

Most important, however, to those involved in automatic assembly equipment, is the necessity to recognize that unless the physical characteristics and features of dimension, symmetry, or substantial asymmetry resulting in gravitational unbalance are also functionally critical, quality control statements as to part quality have little bearing on probable assembly system efficiency.

If possible, assembly machines should be designed to utilize functionally critical dimensions if they do not compromise optimum machine design.

Unfortunately, critical functional features are often internal or recessed making them unsuitable for selection or orientation purposes in part feeders. Other nonfunctional characteristics must then be selected and used for orientation purposes.

Quality Specifications for Assembly Features

All involved in manufacturing are aware that definition of components is made by shape and dimension. The dimensions are by necessity held to tolerances dictated by product function and by the need for interchangeable manufacturing. When Eli Whitney established the principles of interchangeable manufacturing, assembly (albeit by hand) was a vital consideration. For this reason, product designers feel that they are unfairly criticized for failing to consider assembly in their design.

Assembly system designers are concerned, however, not about functionality or interchangeability of components, but the features of shape and dimension that make a part easy or difficult to feed, orient, or select automatically, particularly if these parts come to the assembly system in bulk condition. Once functional requirements are specified by the product design team, specifications must also be established on those features that will determine the capability of parts to be handled automatically. This usually cannot be done on the drawing board and probably will have to be deferred until experimental work is possible with initial product models. A very rapid growth in computer simulation capability may prove extremely helpful in this area.

Since all designers are schooled to recognize that component cost is proportional to tight production tolerances, there is a sensitivity to adding further tighter dimensional specifications, particularly where they are nonfunctional to the product. Product designers can remove much of the problem by recognizing that full symmetry,

substantial asymmetry, or adequate locating features "handles" usually will reduce the necessity of holding close tolerances on components. *The necessity to hold tight nonfunctional tolerances to facilitate component part feeding is usually indicative of failure to provide adequate asymmetry in component design.* In other words, proper design of components to be assembled automatically can often remove the necessity to hold close tolerances in fabrication.

Problems in Matched Assemblies

Increasingly, the search for higher product performance particularly in automotive components requires a matching of component parts to obtain the desired performance characteristics. Fuel injectors, bi-metal-actuated thermostats, and precision bearings are typical examples. Dr. S. S. N. Murty, in an ASME paper, very concisely outlined the problems of matched assembly when he stated the following:

"In the design of part tolerances in precision assembly, *functional tolerance on the assembly may not be economically obtained with parts produced even by processes of highest available capability. The resort to selective assembly is the only viable approach.* In this process the mating parts are machined to the available process tolerances which may be larger than the specified part tolerances. The individual units of a part are classified into suitable number of groups, based on equal division of its process tolerance. The part required to be mated is also classified exactly in the same manner. Corresponding groups of parts will then be assembled to obtain close tolerances on the functional dimension. If the parts in the matching groups are not equal in number, some of the parts will remain unused, thus resulting in mismatch. If the group limits (the maximum and minimum material limits of a group) are altered to minimize mismatch, the mean fit and range of fits of the assemblies become different from group to group. This leads to nonuniformity in the quality of assemblies. The three necessary conditions for an efficient selective assembly process are (i) the corresponding groups must contain as many internal parts as external parts, (ii) range of fit must be same for all groups, and (iii) average interference (or clearance) must be the same for all groups. Any method which satisfies all these conditions will be the best method."

If matched assemblies are required, the basic problem that must be overcome before any mechanized assembly is practical is a realistic evaluation of the possibilities that there will always be a sufficient supply of components with matching characteristics available at the assembly machine. Too often this assumption is not realized, and beautifully designed assembly systems fail because there was not a realistic evaluation of probable parts availability.

Having looked at part quality definitions, it is necessary to refresh our memory on the nature of the sigma curve distribution of any specific part quality, control limits, and applicable mathematical laws concerning probability and possibility.

If we make a sufficient number of parts and examine and plot any one specific characteristic on a graph, we should get a traditional sigma curve as shown in Figure 2-12. If the majority of parts fall within established control limits and the peak of the curve coincides with the mean dimension of the tolerance range, the parts are considered under control. If a significant portion of the parts fall outside the minimum and maximum control limits, the component fabricating process does not have sufficient capability. A decision must then be made to upgrade the process or, alternatively, to relax the controls. If the curve would fit within the range of the control limits, but does not properly coincide with the limits, the fabricating process is capable but requires adjustment.

Do not make the easy and incredibly common error that a sigma curve represents all of the characteristics of any component part; it represents only the distribution of results on that one characteristic being studied. Any given part can have a multitude of characteristics, any one of which would fall on a different place of a sigma curve for that specific characteristic.

In addition to the distribution of tolerances represented by sigma curves, the product designer must recognize the impact of algebraic

laws of probability and possibility on machine output. Since these considerations are more closely aligned with control system design, they are discussed in Chapter 4.

A closely allied problem to part matching was outlined by Dr. M. F. Spotts in his ASME paper, "Probability Theory for Assemblies with Parts Piece Errors Concentrated Near End of Tolerance Limit." The paper is so short that it is worth repeating it in its entirety.

Piece parts can have many different distributions for their errors and a typical example is shown in Fig. 2-13a. There are n parts with mean $\overline{X}$ and tolerance $\pm u_n$ to be assembled end-to-end. The errors in the parts are assumed to have central normal distributions as shown in Fig. 2-13a. Subscript n denotes that the distribution of errors under discussion is normal or Gaussian.

Probability theory shows that the natural tolerance, u_a, for the normal assembly curve for normally distributed piece part errors is given by the following equation:

$$u_a = (\Sigma u_n^2)^{1/2} \qquad (1)$$

When the tolerances, u_n, for all the parts of the assembly are equal, Σu_n^2 is equal to nu_n^2 and the foregoing equation becomes

$$u_a = n^{1/2}u_n \qquad (2)$$

This is the natural tolerance of the anticipated assembly curve for the parts dimensioned in Fig. 2-13a and is shown in Fig. 2-13c.

Suppose, however, the parts are produced under conditions where the variations are less than those specified and at the same time the mean of the production is located to one side as shown in Fig. 2-13b. The errors in the parts as produced are within the specified limits, and the parts would normally be considered acceptable. The question arises as to how the assembly curve for the actual parts will compare with the anticipated assembly curve of Fig. 2-13c.

Because of the shift, $0.4u_n$, of the piece part mean in Fig. 2-13b, the mean of the assembly

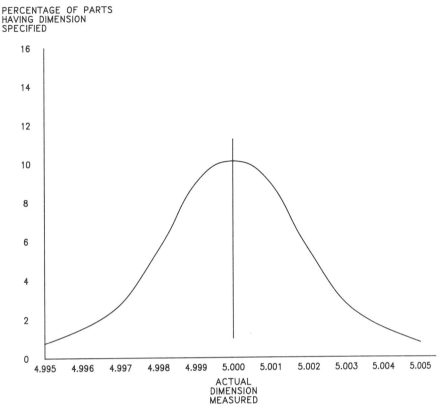

PERCENTAGE OF PARTS
HAVING DIMENSION
SPECIFIED

ACTUAL
DIMENSION
MEASURED

**Fig. 2-12. Sigma curve distribution of parts when plotted against a specific tol-
eranced feature.**

curve will have shifted to the left an amount n times as great or $0.4nu_n$, as shown in Fig. 2-13d. By Eq. (2), the natural tolerance of the assembly curve will be:

$$u_a = n^{1/2} \times 0.6u_n = 0.6n^{1/2}\, u_n$$

The standard deviation, σ_a, for the assembly curve is one-third this amount, as indicated.

The assembly curve extends a distance to the left of the end, B, of the anticipated assembly curve. The shaded area, A, can be considered the proportion of underlength assemblies. The value of A can be found in the probability Table accompanying this article [Fig. 2-13e]. To use the table, a nondimensional coordinate, t, is required which is equal to distance CD divided by the standard deviation, σ_a. Distance CD is equal to $n^{1/2}u_n$, minus $0.4nu_n$ or $(n^{1/2} - 0.4\, n)\, u_n$.

Then

$$t = \frac{(n^{1/2} - 0.4n)u_n}{a}$$

$$= \frac{(n^{1/2} - 0.4n)u_n}{0.6n^{1/2}u_n/3} = 5 - 2n^{1/2} \qquad (3)$$

The value of n can be substituted in this equation to obtain t, which is then applied to the table to get the proportionate area A.

Example: Find the value of A for an assembly consisting of four parts when the actual production is like Fig. 2-13b.

Solution: $n = 4$.

By Eq. (2), $t = 5 - 2\sqrt{4} = 1.00$. By Table 1, $A = 0.159$. Thus, 15.9 percent of the assemblies will be shorter than expected.

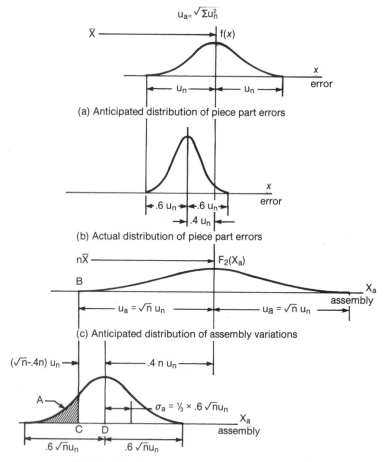

(a) Anticipated distribution of piece part errors

(b) Actual distribution of piece part errors

(c) Anticipated distribution of assembly variations

(d) Actual distribution of assembly variations

(e) — Areas Under the Normal Curve
Proportion of Total Area Between Limits t and ∞

t	A_t^∞	t	A_t^∞	t	A_t^∞
0	0.5000	1.0	0.1587	2.0	0.0228
0.1	0.4602	1.1	0.1357	2.1	0.0179
0.2	0.4207	1.2	0.1151	2.2	0.0139
0.3	0.3821	1.3	0.0968	2.3	0.0107
0.4	0.3446	1.4	0.0808	2.4	0.0082
0.5	0.3085	1.5	0.0668	2.5	0.0062
0.6	0.2743	1.6	0.0548	2.6	0.0047
0.7	0.2420	1.7	0.0446	2.7	0.0035
0.8	0.2119	1.8	0.0359	2.8	0.0026
0.9	0.1841	1.9	0.0287	2.9	0.0019
1.0	0.1587	2.0	0.0228	3.0	0.00135

Fig. 2-13. Out-of-tolerance assemblies owing to excessive part piece tolerance. (a) Anticipated distribution of part piece errors. (b) Actual distribution of parts piece errors. (c) Anticipated distribution of assembly variations. (d) Actual distribution of assembly variations. (e) Area under the normal curve proportion of the total area between limits t and ∞.

The proportion of underlength assemblies increases as n increases since the mean of the actual assembly curve in Fig. 2-13d moves to the left as n becomes larger. Thus, for n equal to 7, over one-half the assemblies will be underlength.

The ill effects described in the foregoing come about when the quality of production is good, less spread for the piece part errors, combined with a shift in the mean. The foregoing results assumed that all the parts of the assembly are so affected as might occur for parts made from identical tools or for the same machine setup. Should only one, or a few, of the parts of an assembly be so affected the proportion of out-of-tolerance assemblies may be much less. A development similar to the foregoing can be easily made for other offsets of the mean and reduced piece part tolerances as well as for other forms of distributions of the piece part errors.

The situation of Fig. 2-13a illustrates the harmful effects that can result from specifying tolerances greatly in excess of the production capabilities. Excessive tolerances should be avoided for no other reason than to prevent the mean from shifting, which is the principal cause of out-of-tolerance assemblies. *The bad effects are more likely to occur in assemblies consisting of a relatively large number of parts.*

There is a wealth of theoretical study on assembly in the proceedings of the Design Engineering Division of ASME. In many cases, tolerances can be opened up and still provide for optimal performance. Fletcher Eaton's ASME paper 75-DE-21, "Computer Modeling of a Three Dimensional Assembly" is particularly interesting. *Automatic assembly does not necessarily mean tighter tolerances, instead it means realistic tolerances rigorously adhered to.*

Features for Automatic Inspection

One of the most attractive features of automatic assembly is the possibility to monitor on an ongoing basis the quality of the evolving assembly as it proceeds through the assembly system. This ability, however, is conditional on designers providing in their design the necessary access for probes and inspection devices in their overall product design.

A most frustrating problem in assembly system design is verification of the presence and position of small parts where the inspected dimension may be less than the tolerance on the stack-up of previously fed parts. Every effort should be made in product design to provide internal reference surfaces for automatic verification of part presence and position, preferably by mechanical contact. Descriptions of various gauging techniques may be found in Chapter 4.

PRODUCTION AND PURCHASING TEAMWORK

In General Electric's seminar activity in training people to design products to facilitate assembly, great emphasis is placed on management's role in designing for efficient assembly, and specifically in four areas outlined below:

Product Planning
  Define Assembly Relationship To:
    Product Performance
    Customer Needs
    Production Economics
    Development Plan
    Product Structuring
    Capital Investment
    Business Strategy
Conceptual Design
  Surface The Assembly Issues:
    Design Alternatives Evaluated?
    Adequate Planning?
      Funding
      Prototypes
      Tooling
    Multifunctional
    Integration
Product Design
  Review and Evaluate:
    Tradeoffs/Strategy Design
    Compatible With Mode of Assembly
    Multifunctional Integration
Manufacturing Planning
  Coordinated Approach To:
    Resource Planning
    Procurement Schedule
    Drawing Review
    Process Planning
    Problem Solving

It is obvious that some managerial input must be involved in every stage of product design.

Integrators of assembly systems have many areas of product design concern that fall within the scope of general management. If we can assume that product design has given full recognition to the various aspects that make for efficient mechanized assembly and that these are correctly outlined on part drawings and specification sheets, there still remains the problems of ensuring that in-house production and outside procurement of components match the specifications.

The ideal situation is where mechanized assembly and component fabrication are done under the same line management in the same location. The ideal is a work cell plant configuration. This ideal situation has until lately rarely existed. Many assembly plants have purchased most or all of their components from either sister plants or outside vendors, particularly when management places great stress on core competency.

The most determined attitudes regarding part quality and shipping methods often collapse in the face of production schedule crises.

The key areas that should be considered are those which ensure that Production, Production Control, Purchasing, and Quality Control are aware of the impact on assembly machine productivity when part quality, cleanliness, and availability fall below specified norms. Where purchased parts are used in automatic assembly, every effort should be made to receive and inspect incoming parts in sufficient time to replace defective lots without disrupting production schedules.

SUMMARY

An awareness of the importance of proper design to facilitate mechanized assembly has grown rapidly in both the business world and in academic circles. Many feel, with much justification,

that it is the most profitable and least costly way to improve assembly efficiency.

This awareness has led to a search for basic principles that can be applied to each design to achieve optimum solutions. The pioneer work of Dr. Boothroyd was followed by many industry driven programs such as the brilliant work done by Hitachi and General Electric. For the foreseeable future, however, such universal design principles are often difficult to apply in specific projects. Software developments attempt to incorporate these design principles to facilitate assembly into the initial product design stages.

The best possible approach is to come up with designs that are functional and marketable and then critique these designs in the light of assembly requirements. There must be a breakdown of the isolation between product design and manufacturing engineers. *There can be no optimal design for assembly without this design being correlated to a specific assembly system. A good design for one assembly system may be a mediocre design for another system. There is a synergism in the proper matching of product design and assembly system.*

REFERENCES

1. R. W. Bolz, *Production Processes, 5th ed.,* Industrial Press, New York, 1981.
2. G. Boothroyd, *Assembly Automation and Product Design,* Marcel Dekker, New York, 1992.
3. F. Eaton, "Computer Modeling of a Three Dimensional Assembly," ASME, New York, 1975, 75-DE-21.
4. A. J. LaRue, "Some Quality Considerations in the Design of Automatic Assembly Machines," *ASTME,* Volume 60, Book 1, Paper 263, 1960.
5. S. S. N. Murty, "Selective Assembly—Its Analysis and Application," *Proceedings of 4th International Conference on Production Engineering,* Japan Society of Precision Engineering, Tokyo, 1980, pp. 913–918.
6. M. F. Spotts, "Probability Theory for Assemblies with Parts Piece Errors Concentrated Near End of Tolerance Limit," ASME, New York, 1975, 75-DE-1.

Chapter 3
Selecting the Assembly Machine System

INTRODUCTION

Before it is possible to get into any specific details of tool design for assembly operations, it is necessary to choose an assembly system approach that best matches the job requirements. Logically, *work station and fixture design can only follow assembly machine chassis or assembly machine system selection.*

Index accuracy, station integration, line balancing requirements, and index-dwell ratios are among many factors which will have a major bearing on gross machine capability. From the operational (and financial) viewpoint, however, it is the realized net production of the system, not the theoretical gross capacity, that counts. Net production over the life of the assembly system or product will be determined by system durability, reliability, ease of maintenance, availability and interchangeability of spare parts and quick and safe access for machine operators to clear inevitable jams. Of increasing importance is the chosen basic system's ability to accept running product changes without significant downtime or extensive rebuilding.

Many factors will determine system and basic chassis selection. There is one element, however, that may reduce the user's choices of machine chassis. Most established system builders have standardized on one or more basic machine systems configurations or integral chassis suitable for a wide range of applications. If experience, delivery, or compatibility with existing equipment dictate the choice of a specific system builder, the choice of an assembly system or chassis may quickly become restricted. It is a case of "love me, love my dog."

Mechanically integral or modular assembly machines offer many operating advantages, not the least of which are energy savings, field-proven equipment and simple system synchronization.

In any case, final chassis selection is generally a compromise. It is well to recognize all the factors in deciding your course.

TYPES OF INDEX

In any assembly system, the basic fixture or pallet transfer indexing mechanism is one of the most vital system elements. The method selected for moving the fixture or pallet together with the developing assembly from one work station to another will determine to a great extent the type of control system used, the mechanical design of the work holding fixture, and both the design and the method of actuation of the individual transfer and inspection systems employed. Tooling costs, machine efficiency, and realized (net) production rates will be directly related to the type of indexing mechanism used.

From an engineering standpoint, the choice of indexing mechanism used is often dictated by the number and size of parts to be assembled, the types of joining operations to be performed, and the production rate required. From an economic standpoint, the ideal system and the total potential savings for any specific assembly operation may be compromised by limited production rates, limited model runs, and alternative low-cost production methods—more specifically, cheap labor. The system may have to provide for ergonomic coupling of machine operators to automatic stations.

The function of the index mechanism, upon which assembly tooling for synchronous assembly systems is mounted, is to transfer the work holding fixtures and developing assemblies before transfer and inspection stations. Generally, transferal refers to intermittent motion of the work holding fixtures and developing the assembly to successive work stations where additional components of the assembly are added, inspected, joined, or the final assembly ejected. While the majority of assembly machines are so actuated, two types of assembly jobs often can be handled simply and economically without such intermittent motion.

Single-Station Machines

The first situation of this type occurs when the assembly work is performed at a single-station machine. The machine shown in Figure 3-1 is typical of such a unit. Here, two, three or more parts are fed to a single assembly station and are assembled and ejected in the same station. Such machines may drive many components into a large housing at one time. Stud inserting on engine blocks is typically done this way. Single-station machines can be either mechanically or fluid power operated. The use of commercially available pneumatic or hydraulic components or systems in simple machines helps to keep assembly systems design and fabrication costs down. The relative simplicity of air circuits gives the ma-

chine a fairly high reliability factor. Inspection by electric eye or other probes for presence of parts may aid in ensuring that only completed assemblies leave the machine.

Continuous-Motion Machines

The second type of mechanized assembly work not requiring any intermittent motion mechanism for fixture transfer is that performed on a continuous-motion machine. Here, parts are assembled in much the same way as brass rings are seized on a merry-go-round. Early components in the assembly sequence being produced such as pins or shafts, traveling in either a straight line or circular path, are used to pull subsequent component parts from detented chute rails (Figure 3-2). This continuous-motion concept is commonly used in packaging and filling machinery. A quite different example of a continuous-motion operation is the assembly of "O" rings to plates in a single, common vibratory feeder, as illustrated in Figure 3-3.

Obviously, this method of assembly without multiplanar transfer units is suited mostly to parts of symmetrical configuration, such as rings, cylinders, and washers which require no specific radial orientation. Because of the time saving achieved by eliminating index time and the high production rates that are possible with continuous-motion arrangements, the assembly of non-symmetrical parts requiring more-sophisticated transfer devices has often been attempted on a continuous-motion chassis. Results have varied widely. Transfer devices on such machines must move in lateral or radial synchronization with the work-holding fixture. Joining operations become extremely difficult, particularly when line balancing is required. The complexity of required station tooling often can result in more lost production than is caused by the lost time of intermittent motion. The development time and cost of such tooling may also offset anticipated savings to be achieved through potentially higher production. This type of approach is usually suited only for extremely high production and serial production of identical machines. Most companies developing assembly systems for specific one-of-a-kind application no longer attempt continuous motion machinery.

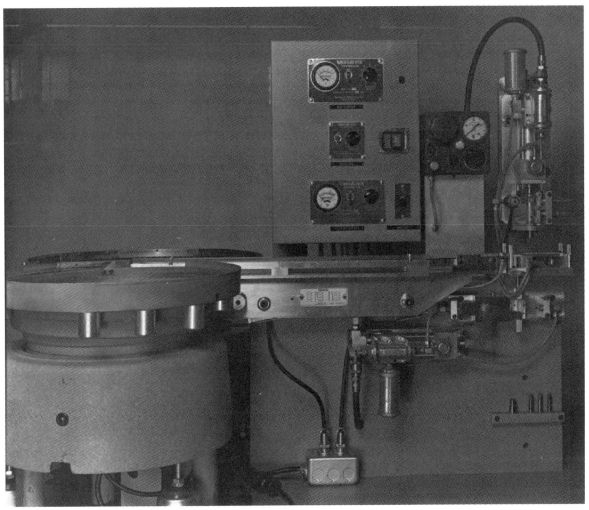

Fig. 3-1. A single station assembly machine. Bonnet nuts and packing are fed down separate tracks. When photocells sense the presence of both parts, an air cylinder inserts the packing into the nut.

Intermittent Index

The majority of assembly machines for small and medium size products are built around various forms of intermittent motion assembly systems. While the variations that such devices can take form a tribute to the ingenuity, or perhaps, the perversity of man, there are four basic methods of producing an intermittent motion mechanically. In each case a circular or reciprocal motion is converted into an intermittent unidirectional linear or radial movement. Each total machine cycle is composed of an index period and dwell period.

Fluid Power. A growing awareness of the true cost and environmental hazards of using fluid power, air or hydraulic to actuate industrial machinery has made many people reluctant to use it as a prime mover. Compressed air is expensive to produce and to distribute efficiently. Escaping air is a major source of industrial noise. Containment of this noise may increase capital equipment costs and reduce accessibility for maintenance and operation. Hydraulic power generally requires an external coolant source such as city water or closed-loop systems using refrigeration or evaporative towers. These envi-

Fig. 3-2. Continuous motion of the indexing dial wipes away cylindrical housing, solder rings and washers to form a diode heat sink on this small dial-type machine.

Fig. 3-3. Two separate parts are placed in bowls on common base which automatically combines plates and "O" rings.

ronmentally sensitive energy, water, and noise considerations must be taken into account when considering fluid power.

Fluid power produces linear or radial intermittent motion by extending and retracting a cylinder rod to drive a pawl against a cylindrical ratchet for radial motion or a series of pawls to produce intermittent motion in a linear path. (Figure 3-4.)

The advantages of fluid power include low initial component cost, broad availability of replacement components and maintenance skills, and additionally provides ease of establishing varying index-well ratios. The disadvantages include a possible need for upgrading plant compressed air facilities and relatively high maintenance costs found in high cyclic applications.

Ratchet and Pawl. This mechanical system is perhaps the oldest known form of intermittent

motion mechanism and is used almost exclusively on rotary machines. It is commonly used to actuate air-operated indexing tables. Mechanically actuated ratchet and pawl index mechanisms (Figure 3-5) have the advantage of rapid changeover from one index ratio to another. If the ratchets and cams are easily accessible, maintenance work is comparatively simple. The ratio of index to dwell time is usually favorable. Also, it is possible to move the dial freely in one direction without operating the machine under power, since ratchet and pawl systems are not usually mechanically interlocked as is the case with the Geneva or crossover cam types of index. This freedom of dial movement is often a great aid to operators or mechanics in setting up the machine and clearing jams.

The disadvantage of this type of index is the possibility, particularly when fixtures are heavy, to have the indexed work fixture holding dial overtravel in the direction of rotation when the pawl or locking pin are disengaged for return to the next ratchet tooth. If the pawl is actuated by a crank, severe acceleration and deceleration characteristics occur. If, however, the indexing pawl is cam actuated, a desired acceleration and deceleration curve usually can be achieved.

Geneva Mechanisms. The Geneva, or star wheel system (Fig. 3-6), is an early attempt to

Fig. 3-4. Fluid power reversal on a cylinder causes intermittent radial or linear motion.

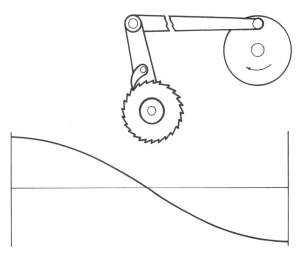

Fig. 3-5. Ratchet and pawl mechanisms have been used to produce intermittent motion for centuries.

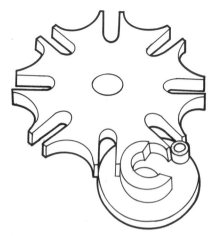

Fig. 3-6. An eight position Geneva mechanism. One rotation of the small wheel will cause a 45° rotation of the large wheel which is then locked until the next rotation of the small wheel.

combine the mechanisms used for indexing and for locking the fixture carrier in position during the dwell portion of the cycle without any overtravel. While the Geneva mechanism is still widely used on tooling stations, it is no longer as popular as it once was for basic machine drives. The relatively large portion of the machine cycle time consumed in indexing and the severe acceleration/deceleration characteristics of this type of drive mechanism have been severe operational weaknesses.

Crossover Cams. The most common form of mechanical intermittent motion device in use today is the crossover cam shown in Figure 3-7. Simple in construction, it has excellent acceleration and deceleration characteristics. Its kinematic properties are particularly adapted to high-speed operations. The crossover cam can be directly fastened either to a dial for use on rotary machines or to drive sprockets or drive wheels used to move pallets fastened to chain or belts on in-line machines. (Figure 3-8.) The chief mainte-

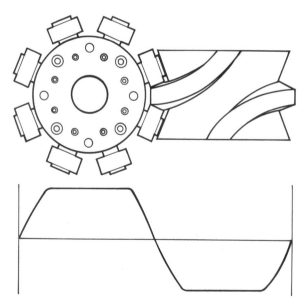

Fig. 3-7. A crossover cam indexer. A series of rollers in continuous motion engage a cam wheel resulting in intermittent motion.

Fig. 3-8. The crossover cam indexer used in this linear machine is driven by a speed reducer. The intermittent output is coupled to the drive wheel of the machine through a sensitive clutch which protects against fixture damage.

nance problem on these types of machines has been lack of accessibility to the crossover cam and rollers on dial-type machines. More recent linear assembly machine designs using crossover cams have attempted to correct this accessibility problem by locating the index device outside the tooling area of the machine.

Rectangular-Motion Devices. Rectangular-motion or walking-beam-type systems are commonly used to actuate in-line machines. (Figure 3-9.) These units move either pallets or the main workpiece itself. Such systems can be extremely versatile. Work-holding pallets can be removed readily from the machine for subsequent operations or the same pallet can be used on both fabricating and assembly machines. When it is possible to use the main housing of an assembly as a fixture without a pallet, tooling costs are significantly reduced (Figure 3-10).

Other Indexing Methods. In addition to mechanical methods for obtaining intermittent motion, some assembly systems have been built with an electrically controlled or activated indexing mechanism. Such units may couple the drive motors directly to gearing or chain sprockets and obtain an intermittent motion by alternately starting and stopping the drive motor. Here, the main design problems are index accuracy and reliability of electrical controls. The most common system is to couple the main drive motor to the intermittent motion device or index drive wheel through a clutch, eliminating the need to stop and start the main drive motor. While this approach gives a degree of flexibility in varying index to dwell ratios, it precludes the continuous synchronized rotation of main cam shafts, which are key elements of integral assembly machine design discussed later in this chapter.

Relationship of Index Time to Dwell Time. From a production standpoint, the machine time spent in indexing fixtures from one assembly station to another may be lost time. Productive work, such as component insertion and joining operations on the actual assembly, is done, for the most part, on the dwell portion of the index cycle.

Index time is at best a necessary evil. On me-

Fig. 3-9. A typical walking beam machine. Intermittent index is caused by a reciprocating bar with adjustable pitch features.

Fig. 3-10. Rectangular switch bases are carried along a stationary track without fixtures by a reciprocating comb and are held in position at work stations by a locking cob.

chanically operated intermittent-motion devices, a continuous-motion power input is converted into an output of alternate fixture motion and rest. Index time is that portion of a single assembly system cycle used to produce the fixture transfer function. Once fixture locking pins are in place, and a time deduction made for tool advance and withdrawal, the balance of the machine cycle is available for the assembly functions, such as part transferal, inspection, or joining. Roughly 50% of total system cycle time is available for work on the product.

Many assembly operations such as component parts insertion are of themselves practically instantaneous. Tooling need only advance to a fixed position and return. However, some operations require specific time periods. Welding, screw driving, gauging, machining, and spin riveting are among those operations that require some time interval for completion. The speed of the main power input shaft is set so that the dwell portion of the machine cycle is equivalent to the time required to advance the tooling, perform the assembly, joining or testing function, and withdraw the tooling. *Index time, when of a fixed ratio to dwell time, is determined by the time required by the work portion of the total cycle.*

Thus, the portion of the machine cycle consumed in indexing is an important design and purchase consideration. In many cases, index time is a function of the number of assembly joining or testing stations and resulting possible inertial loads on the driven member of the index-ing mechanism. However, this is not always true for ratchet and pawl units, which usually require the least percentage of the total input cycle for indexing.

Unfortunately, terminology about index times can be misleading. It is especially important to know if the index time specified is only that por-tion required for the indexing function, or if it also includes the time required for advancing and withdrawing the tooling.

When machine cycle rate is reduced to per-mit a sufficiently long dwell period for joining or gauging operations, and the intermittent mech-anism input speed is constant, the assembly system chassis may be indexing far below its mechanical capacity and, therefore, losing If programmed motor drives or motors with high inertia rotors and integral clutches are used, the relationship between index and dwell events can possibly be altered until each consumes the op-timum time. This is usually only practical on very slow (long cycle time) systems. Significant ad-vances in motor controls may dramatically alter this constraint.

A complete understanding of index-dwell ratios becomes very important when one critical oper-ation in the total assembly process has a signifi-cant time imbalance with other operations. Some ingenuity is required to achieve overall system line balancing, thus producing the least total cy-cle time. Some line balancing approaches are dis-cussed later in this chapter.

Programmed Drives. Integration of the index function to the insertion, joining and gauging op-erations of work stations can be done by me-chanical integration (e.g., by cam shafts, sprock-ets, or gearing) or by electromechanical and electropneumatic coupling.

In an age oriented toward ever new electronic marvels, it is easy to ignore mechanical simplic-ity and make all machine system sequencing de-pendent on electronic control. The chapter on controls, which follows, reviews commercially available control choices. It is sufficient in this chapter to point out that there will always be some degree of necessary interfacing between mechanical and pneumo-electrical devices re-quired in almost every assembly system. In mak-ing a selection of basic machine chassis, it is es-sential to determine how the mechanically in-dexing chassis and its main camshafts can be most efficiently coupled to increasingly complex machine control systems. (Figure 3-11.)

Index Accuracy. In spending several decades negotiating the conceptual development of pro-posed assembly machinery, the author has found that a most common question of a potential cus-tomer is, "What is the index accuracy of the ma-chine?" This is a most intriguing question; and any answer must be determined by the form and objectivity of the measurement.

Usually people are concerned, and rightly so, about the output accuracy of the basic work hold-ing fixture indexing device. For instance, what is the absolute index accuracy of each tooth of a drive ratchet? In the case of crossover cams, the problem is compounded because of manufactur-ing deviations in each cam path and the radial location of each driven roller together with its in-dividual eccentricity deviations.

Since in most normal applications the inter-mittent motion device index radius is smaller

Fig. 3-11. Output of a magnetic shaft encoder dri-ven from the index programmable controller with the mechanical movements of the machine.

than the work-holder fixture drive mechanism radius, very small errors in the intermitter are significantly compounded in the work holding fixture rotary table or linear fixture transport system. It is often necessary, therefore, when fixture index accuracy relative to transfer station location is critical, to utilize fixture locating shot pins in the work holding table or fixture drive wheel, preferably at the largest possible radius, to obtain the best possible fixture locational accuracy. When this is done, some form of dimensional compliance will be required between the indexer and the work table.

What the user really wants to know, however, is "What is the index accuracy of the workpiece?" For instance, assuming absolute accuracy in the indexing system, we still have the problems of the relative location of work holding fixtures to the work table pallets or drive wheel. Additionally, we must have some nest clearance in the fixtures to allow for assembly component part variation. We then must consider cavity location in the workpiece components relative to the external features used for staging. Finally, we have the problems of transfer device accuracy as described in Chapter 5.

Index accuracy is an essential consideration of machine index system design, but it is not, in itself, sufficient to ensure totally proper alignment of already inserted components to receive additional components. All four elements of overall machine index accuracy, fixture nest accuracy, uniformity of assembly components and transfer unit insertion locational accuracy relative to work holding fixtures must be considered jointly.

Selecting an Indexing System. As a rational basis for selecting a specific form of indexing mechanism or for choosing between the use of a series of individually powered work stations or an integral assembly machine, the following factors must be examined. The first factor is the minimum achieved production rate that can be accepted. Tables 3-1 and 3-2 give both annual and hourly production at various net levels for specific machine cycle rates. *For cyclical industries that have strong seasonal peak loads, the highest required hourly production requirements are more important than annual production capabilities.*

Industries, however, producing high volume consumer goods, particularly those of a disposable nature, usually have both uniform production requirements and excellent design stability. Machine costs can be amortized over longer periods. Such high volume production combined with design stability could possibly justify the design of a completely special machine and, often, justify prototype machine development prior to actual fabrication of production machinery. Low salvage value of the prototype assembly system and subsequent machinery would not be critical in this instance.

For industries producing durable goods where annual volume is in the low millions but where style or competitive considerations set definite or potential limitations to the length of model runs, prototype assembly system development is impractical, and the use of some standard integral assembly machine chassis which can be easily modified or retooled is clearly indicated. Not only is salvage value of such a standard system high, but the components of such standard machines are usually stocked by the builder. Thus, delivery of a basic, untooled machine or maintenance parts is usually not a problem. In addition, builders of standardized integral assembly machines have standardized transfer units and inspection probes that further reduce system development time.

Any of the various intermittent-motion devices previously described are capable of operating easily up to 60–70 assemblies per minute. If machine speeds beyond this cyclic rate are required to meet production requirements, then tooling considerations such as gauging and control response times and parts feeder outputs must be analyzed to determine if they are adequate for the higher cyclic speeds. If they are not, multiple machine lines or multiple system tooling may be required to meet production requirements.

When component transfer or gauging stations of the assembly machine are pushed by production demands to a cyclic rate where their efficiency becomes marginal and output is affected, the possibility of increasing the number of assemblies processed at each cycle of the machine should be examined. Here parts quality and probable station efficiency are the determinants of the

TABLE 3-1. ANNUAL MACHINE PRODUCTION

| Machine rate (cycles/min) | When ratio (%) of net production to gross production is: | | | |
	100%	83% Annual Production*	75% (units) is:	60%
10	,200,000	1,000,000	900,000	720,000
20	2,400,000	2,000,000	1,800,000	1,440,000
30	3,600,000	3,000,000	2,700,000	2,160,000
40	4,800,000	4,000,000	3,600,000	2,880,000
50	6,000,000	5,000,000	4,500,000	3,600,000
60	7,200,000	6,000,000	5,400,000	4,320,000
70	8,400,000	7,000,000	6,300,000	5,040,000
80	9,600,000	8,000,000	7,200,000	5,760,000
90	10,800,000	9,000,000	8,100,000	6,480,000
100	12,000,000	10,000,000	9,000,000	7,200,000
110	13,200,000	11,000,000	9,900,000	7,920,000
120	14,400,000	12,000,000	10,800,000	8,640,000

*Annual production capability for a machine completing one assembly at each cycle. Data is based on a single shift, 40 hour week. Production for machines processing two or three assemblies at a time, or for multiple shift operation is obtained by multiplying production shown by that factor. This table assumes a 480 minute shift and must be factored for shorter work days.

feasibility of manufacturing two or three assemblies per stroke.

One final consideration is whether to handle all production on one high-speed machine, which may be marginal and may require occasional hand line back up, or whether to break up the production over two slower machines with the resultant higher equipment costs.

In an early article in *Automation* Kenneth Treer wrote:

"Many conditions dictate the type of intermittent transfer device best suited for a given application. The following factors should be considered when selecting an intermittent transfer device for a specific application:

Number of production functions required.

Physical size and weight of parts and total assembly.

Weekly, monthly, or yearly volume of parts and hourly production rate desired.

TABLE 3-2. HOURLY MACHINE PRODUCTION

| Machine rate (cycles/min) | When ratio (%) of net production to gross production is: | | | |
	100%	83% Annual Production*	75% (units) is:	60%
10	600	500	450	360
20	1,200	1,000	900	720
30	1,800	1,500	1,350	1,080
40	2,400	2,000	2,000	1,440
50	3,000	2,500	2,250	1,800
60	3,600	3,000	2,700	2,160
70	4,200	3,500	3,150	2,520
80	4,800	4,000	3,600	2,520
90	5,400	4,500	4,050	3,240
100	6,000	5,000	4,500	3,600
110	6,600	5,500	4,950	3,960
120	7,200	6,000	5,000	4,320

*Hourly production capability for a machine completing one assembly at each cycle.

Permissible transfer time as part of total operating cycle.

Accuracy of transfer required to ensure proper location and placement of parts or other station functions.

Inertial loads involved for acceleration and deceleration of transfer mechanism.

Forces to be applied at various stations.

Need for variable speed drives to meet varying production requirements and assembly speeds.

Flexibility of timing transfer device to permit changing of operating cycle.

Type of transfer device drive—pneumatic, hydraulic, electric, or combination of two or more types.

Floor space and service facilities available for installation and operation of machine.

Initial design, build, debug, and installation costs.

Maintenance costs."

This list remains a very good and comprehensive check list in reviewing possible basic system choices.

Power-and-Free Systems

When workpieces, fixtures, or the combination of the two results in heavy weight (e.g., 5 pounds), inertial problems in transfer of parts may point to the choice of a machine system other than one with direct coupling of fixtures to continuous or intermittent motion devices. Heavy product/fixture weight and severe line imbalance, particularly in joining or gauging operations, point to the choice of some form of free pallet transfer machines (Figure 3-12). These are often called power-and-free machines.

In a free pallet machine, pallets are transferred along carrier tracks where they are stopped at specific work stations for parts feeding, gauging or joining operations. When the work at any station is complete, the pallet is released to the fixture transfer tracks. These work holding pallets may contain internal drive motors and move themselves along the tracks. They may be pulled or pushed along the tracks by chains or reciprocating bars. Increasingly, pallets may move along plastic conveyor tracks. The pallet tracks may be arranged in parallel on some parts of the machine so that several identical stations work on long operations concurrently. These identical stations may also be in a series arrangement providing sufficient queuing capability is built into the system to allow for fixture float.

Each manufacturer of standardized free pallet or transfer machines has unique features peculiar to their equipment. Power-and-free machines are designed for relatively slow operations. The larger and more complex pallets and fixtures are quite expensive. If fast pallet transfer occurs between stations, a decelerating device or cushioning mechanism is necessary at each work station to prevent the transferred pallet from slamming into other pallets or into the work-holding station.

In most cases, a power-and-free system will require extensive electronic controls, locking pins to locate fixtures at each work station, and a very large floor area. It is realistic to consider such a machine as a series of free-standing single-station machines with an integrated automatic part transfer and automatic in-process storage system.

The 1990's have seen significant commercial development of large power-and-free assembly systems which greatly reduce inertial loads and power requirements (see Figure 3-13).

Energy Considerations

In an age of increasing environmental awareness, machine selection must factor into any justification study the energy costs of each proposed system. This is particularly true in areas where consumer groups are fighting to eliminate the quantity discounts on energy that industry has enjoyed for years.

The cam-actuated integral assembly machine shown in Figure 3-21 requires only a single three horsepower motor to provide for all necessary machine and station movement. If this same machine were operated by fluid power, it might require a 20 or 30 horsepower compressor to provide equivalent operating power.

Compressed air is a versatile, flexible, but expensive power source for assembly systems. Not only are compressors costly to run, but after-coolers, refrigerators, and dryers add to the energy bill.

Fig. 3-12. Power-and-free machines are a series of single station machines coupled by conveyor tracks in series, parallel, or both, often with operator work station positions.

Sometimes a different choice of system or station designs may utilize energy more efficiently. For example, a machine that could solder by induction energy rather than natural gas flame might save several hundred thousand dollars during the life of the machine.

We have been careless of energy costs in the past. A new sensitivity to the energy costs and a realistic application of these costs to specific projects may make a modern integrated cam actuated machine far more cost competitive than initial quotes might indicate.

WORK PATHS

Perhaps far too much has been said about the so-called conflict between rotary and in-line assembly machines. In many cases the same assembly can be tooled on either type with equal initial success. Most assembly machine builders offer both types of machines.

Among the more important points to be considered in comparing rotary and in-line machines are work-holding fixture costs and visual and manual accessibility for both operating and maintenance personnel. Rotary machines have an advantage where fixture costs are high. Rotary machines by design necessity realize a much

higher utilization of fewer work holding fixtures than in-line machines. On rotary systems, any fixture is in position for immediate reloading once a completed part is ejected and the fixture is checked for an empty condition. On many in-line machines the emptied fixture, after completed parts ejection, must return some distance to the point where the new assembly sequence begins.

In-line machines offer better visual and manual access and can more readily be sized to fit the job requirements and available shop space. These are only a few of many such considerations in determining the overall best work path for any assembly application.

Dial-Type Machines

Dial type or rotary assembly machines can be as simple as a rotary index table (Figure 3-14) with intermittent motion supplied by a variety of the means previously described, or as complex as a large center column machine.

Rotary index tables are commercially available in a variety of sizes, table diameters, and numbers of stops or index positions. Their use in automatic assembly projects should be limited to the simplest jobs and those with relatively short model life.

The basic problem with using a simple com-

Fig. 3-13. Large power-and-free system assembles engine heads automatically.

mercial rotary index table without any tooling platforms or tooling actuators is the necessity of coupling all transfer, joining, and inspection functions to the index cycle of the work fixtures by use of some control system—pneumatic, electropneumatic, or electromechanical—rather than by simple mechanical coupling. Such machine systems are initially relatively inexpensive, but have higher maintenance downtime and repair costs than do mechanically integral machines.

Sophisticated system builders use dial-type machines which include, in addition to the basic index functions, tool-mounting surfaces and power takeoff devices that are capable of mechanically

synchronizing tooling station motion to the dwell and index of the machine.

These dial-type assembly machines may have a knee and column structure (Figure 3-15) or a center column structure. Knee and column construction is usually limited to table diameters under 36 in. (1 m). Such construction usually restricts tooling to approximately 300° of the dial periphery. It permits extensive inverted tooling actuation, since much of the index mechanism is contained in the column of the machine rather than under the indexing dial. This construction also provides extensive horizontal camshaft availability for tooling actuation.

Fig. 3-14. Rotary machines are compact, visually accessible machines ideal for high volumes and simple assemblies.

Center column machines (Figure 3-16) are rotary machines where the indexing dial revolves around a structural center column. This center column contains not only the intermittent motion device, but fixed and reciprocating tool-mounting platforms mechanically synchronized to the index mechanism. Such center column machines usually have indexing dials of 25 to 75 inches (63.5–190.5 cm) in diameter. They can carry relatively heavy fixturing, and the larger center column machines can carry quite heavy subassemblies.

On the positive side, dial or rotary assembly machines offer many attractive features. They usually have, as mentioned earlier, a very high fixture utilization rate. This means that almost every work holding fixture is being utilized for a part transfer, joining, or inspection function. Machine chassis costs are usually significantly lower than linear machines, and it is possible to obtain high index accuracy at modest cost. Factors weighing against the use of rotary machines include not only those inherent in the machine chassis itself, but those which result from tool design weakness or errors introduced or mandated by the initial selection of a rotary chassis.

The nature of a dial or rotary assembly ma-

Fig. 3-15. A knee and column rotary machine with power sources outside the machine dial.

Fig. 3-16. A center column machine usually has one or more vertically moving tooling platforms as well as a stationary center tool mounting platform.

chine when viewed from above is similar to a pie cut into slices, the number of slices equivalent to the fixtures on the machine (Figure 3-17). Each segment grows in size as it gets farther away from the center axis of the machine. Part feeders fit nicely into these triangular segments, since the escapement and track are narrower than the feeder bowl or hopper.

It is possible to cluster a very large number of feeders and related tooling around a small rotary machine. This compactness or density of tooling often leads to severe operational problems. It has been mentioned earlier, and will be repeated over

and over again in subsequent chapters, that *the novice assembly machine designer designs for the assembly function alone, whereas the mature designer designs not only for required transfer, joining or inspection functions, but provides access for and capability to clear jams and faults in the shortest possible time. Freedom of access to clear these jams (a normal occurrence in automatic assembly) will be the major controllable factor determining net production.*

The very nature of rotary assembly machine layout leads to overlooking the truly essential requirement for easy manual access to both fixtures

Fig. 3-17. Each fixture on a dial-type machine has a wedge shaped area available for tooling. Seen in plain view, it is easy to mount vibratory feeders compactly in these wedges.

and work stations for successful operation. Compactness in original rotary system concept layouts can lead to absolute congestion during final design, erection and debugging. It is very common during the duration of the design, build, and debug segments of system development to find it necessary to add additional tooling beyond that originally envisioned. Such compactness leads to further operational problems when vital index mechanisms and control components are buried under work-holding fixtures and tooling stations, making the system a maintenance nightmare.

It is a common error among novice system designers to fail to leave adequate room for inspection probes after each part transfer operation, including ejection. Experienced assembly systems builders attempt to leave at least 15–20% of available fixtures in initial layout and proposal drawings open for machine and product evolution.

When center column machines are used, tooling movement and sequence may be inhibited by the one or two main cam motions available. Tooling is usually actuated by central master cams rather than individual station cams, resulting in compromise of optimal cam configuration for any specific assembly or test function.

The rotary assembly machine will always remain popular for its compactness, modest cost, and control simplicity. Assembly complexity or combined fixture/product weight may indicate other chassis selections.

Linear Machines

There are several fundamentally different types of assembly chassis that move the workpiece in a straight line. One early selection factor is a determination of whether the base of the assembly being manufactured can serve as its own fixture (comparatively rare) and whether the weight and complexity of the work - holding fixture, line balancing requirements, and planes of insertion access indicate direct coupling of the fixture to the intermittent motion device, or whether such fixtures or pallets should be independent of the fixture transport mechanism through some form of power-and-free design.

Carrousel Machines. Most assemblies will require a work-holding fixture to stage the first components of an assembly to receive additional component part pieces and additionally provide access and support for joining and inspection operations. If the limitations of dial-type machines indicate a linear machine system, synchronized carrousel machines may be a good answer.

A synchronous linear assembly system is usually one in which the fixtures or work pallets are permanently fastened to some transport or carrier device such as a chain or band. At each cycle of the main camshaft, all work-holding fixtures advance one station. These pallets may be fastened to some flexible transport material such as chains and steel bands, or the pallets may be joined to each other like sections of a large hinge to form an endless belt.

The main advantage of such linear synchronized machines lies in the main system camshaft or camshafts, which can run parallel to the fixture transfer path and which can be as long as the total job requirements and adequate station reserves indicate. Additionally, such machines require minimal operational or programming control circuitry because of their essentially mechanical design. A typical synchronous in-line machine is illustrated in the following section on integral assembly machinery.

There are several major disadvantages to synchronous in-line machine systems that may be sufficient to indicate the choice of another configuration of an assembly system. As in any synchronous system, there must be a concern when severe line balancing problems created by the required assembly, joining or inspection processes occur. This is discussed later in this chapter. Since the work-holding pallets are transported by linkage directly coupled to drive wheels or gears, the cumulative weight of fixtures/product being assembled presents significant inertial load considerations. Index accuracy on chain-driven or hinged pallets is not very good and usually deteriorates as the machine gets older. The most widely used synchronous linear system utilizes a prestressed steel band to overcome these index accuracy problems.

If the work-holding fixtures are cantilevered from the carrier pallets, both sides of such a system may be used, but this also mandates stringent limitations on the total fixture/workpiece weight.

The initial procurement cost of such linear systems is higher than for an equivalent dial or rotary system. This higher purchase cost will tend to restrict the application of such machines to assemblies with some degree of complexity and with a significant degree of inspection operations.

In general, synchronous linear machines are used where high production is required; fixture product weight is modest; assembly, joining, and inspectional requirements are moderately to highly complex; and line imbalance is low to moderate.

Over-and-Under Machines. Several linear machines are manufactured in an over-under configuration as shown in Figure 3-18. This eliminates the need for cantilevered fixtures and allows for significantly heavier fixture loads. This configuration also provides good access for horizontal insertion from the two sides perpendicular to fixture travel as well as vertical insertion. It has, however, very poor fixture utilization since returning fixtures are upside down. This design prohibits any extensive operations from the in-

Fig. 3-18. Over-and-under design. The advantages of weight carrying ability and tool access from both sides must be weighed against low fixture utilization and the difficulty of working on the fixtures from the inverted position.

verted position. Returning upside-down fixtures may facilitate fixture cleaning, if that is necessary.

Free Pallet Machines. Free pallet assembly systems are transfer machines wherein pallets, each independent of one another, proceed down transfer tracks to individually cycled work stations, which operate on demand when a pallet is properly positioned before the station. These machines are called power-and-free and are called by the industry "nonsynchronous linear machines." They are nonsynchronous in the sense that each work station operates only on demand when one pallet is properly positioned before that station and without regard to the position of other pallets or the condition of other work stations. Their ability to use parallel work paths or multiple queuing work stations are significant aids to line balancing.

Nonsynchronous machines are relatively expensive. They may or may not include quick fix-

ture or pallet return features to return empty pallets to the starting position if they do not form a totally closed loop fixture path.

Such pallets may have internal drive motors or shunt motors to provide their own indexing motion, and may incorporate decelerating devices or brakes to slow them as they enter work stations.

Application of nonsynchronous machines would generally be indicated when one or more of the following situations occur:

- Very heavy fixture/workpiece weight.
- Annual production quantities of 500,000 to 3,000,000 assemblies.
- Multiplane tooling access is required.
- Severe line imbalance conditions exists.
- High levels of on-line rework for product being assembled are expected.

By their nature, such machines require extensive control circuitry, extensive interlocking systems, and, hence, high initial costs. While inherently flexible, the initial costs, long lead times, and high changeover costs limit the use and application of large free pallet systems to major producers.

Walking Beam Machines. Some assemblies are so designed that the housing, base or major component of the assembly is physically stable so that the product can be pushed along a track to receive additional components without being carried in a fixture. This capability significantly reduces machine costs by eliminating or simplifying the need for fixtures (Figure 3-19). In some instances, free pallet systems utilize a walking beam transfer mechanism (Figure 3-20) by using a track along which the assembly is moved, transfer fingers or combs to move the part laterally, and some form of locking device to stabilize the workpiece during part transfer and joining operations.

The advantages of walking beam machines include lower machine chassis costs and significant reductions in fixturing costs. This transfer system has low inertia, and the index pitch can be varied easily to accommodate parts of varying sizes. This system lends itself well to very high-speed machines.

Unfortunately, the percentage of assemblies that can be assembled without fixtures is relatively small. When it is necessary to stabilize the product being assembled in fixtures, most of the competitive advantages of walking beam machines begins to fade. It is necessary to add some form of fixture return mechanism, and maintenance of such free fixtures becomes a real operational concern.

INTEGRAL ASSEMBLY MACHINES

The necessity of complete coordination between the index and work functions of the machine cycle has led to the development of integral assembly machine systems. Instead of having many individually powered stations, all coupled by a complex timing and interlocking control system, the integral assembly machine is a basic chassis that combines operation of the indexing function and actuation of all tooling from one single power input (Figure 3-21). Mechanical coupling of the index section to the tooling section protects tooling from damage due to improper timing, and simplification of pneumatic and elec-

Fig. 3-19. Small rectangular switch bases serve as their own assembly fixtures as they move along a track. The rear bar moves in and out to lock parts in position for insertion. The front bar moves parts laterally for the next operation.

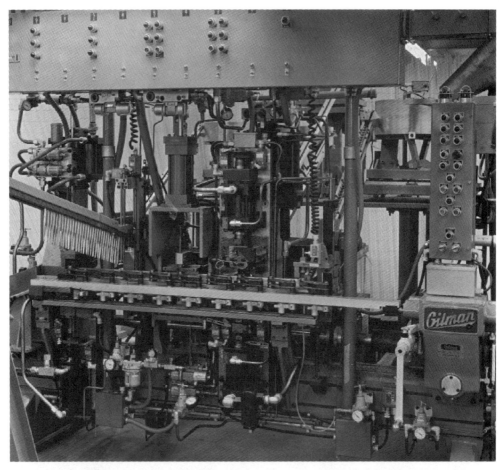

Fig. 3-20. A walking beam machine in which pallets are pushed through the machine.

trical control circuits. Additionally, it reduces potential sources of downtime, increasing machine efficiency.

Integral assembly machines can be classified in several ways. Usually, an assembly machine chassis is classified by describing the path the work follows in traveling from one station to the next. Rotary or dial-type machines are contrasted to in-line or straight line machines. In addition, in-line machines are subdivided into wrap-around and over-and-under machines. Another means of classifying machines is by a method used to actuate both the index and transfer units, that is, whether it is fluid power or mechanical.

The integral assembly machine shown in Figure 3-21 represents one of the more common types in use today. Such commercially available machines have incorporated in their design spe-

cial features that meet certain functional needs that recur repeatedly in assembly work. Historically, these designs reflect the system builder's prior experience in one or more fields, therefore certain designs are more suitable for specific industries than others.

The integral assembly machine represents a very definite philosophical approach to mechanized assembly. It is diametrically opposed to the floor-up design approach to special machinery.

In any assembly project there are four areas that will determine machine productivity:

1. *The basic machine and control system selected.* Does it have the accuracy, durability, and flexibility to meet prolonged production schedules and accept running changes or modifications to the product?

2. *Individual work stations.* Does the work station have the capability of performing the transfer, joining, or inspection function for which it was intended on a high production basis; is it able to cycle millions of times without undue maintenance?

3. *Product design.* Is the product or subassembly itself and its individual components so designed as to facilitate reliable assembly, and permit interface tooling, such as escapements, pick-up fingers, and inspection probes, to be designed in simple durable fashion?

4. *The user environment.* Do the intended operating facility's personnel have the training desire and support capability to operate the machine efficiently, an ability dependent on full teamwork in Design, Production Engineering, Production Control, Purchasing, Sales, and Quality Control?

The special machine builder designing a one-of-a-kind, floor-up unique response to a specific assembly problem must face up to all of these four areas at one time. If the builder is continuously involved in special assembly machine development, he or she faces three internal problems. They will be torn between the normal problems of machine development and the many unique problems of each assembly and its potential manufacturing environment. They will be reinventing the wheel, designing new stations each similar but not identical to another station used elsewhere. They will, of necessity, be debugging the machine system concurrently with the tooling. Each machine remains a prototype. Delays and rework, so normal to prototype development, are inevitable.

Most major builders and many competitive regional or captive builders have faced up to these problems through the development or utilization of integral standardized assembly machines. Such machines have been designed in the light of the builders' past experience and in view of perceived market requirements for the specific industries they wish to serve.

Many of the first attempts at standardized integral machines, both rotary and linear, centered on light bulb manufacturing operations. It was not until the period of the 1950's and early 1960's that full-scale universal assembly machines were developed.

Fig. 3-21. An integral linear machine. Among the features are (a) isolation of index from tooling; (b) upper and lower camshaft availability for tooling actuation; (c) provision for inexpensive inspection device mounting; (d) standardized tooling stations; (e) modular expandable construction.

An integral assembly machine should include the following features:

- A means of accurately indexing the workpiece from station to station.
- A means of directly coupling the mechanical components of the machine to the electrical and pneumatic controls for the machine.
- Some mechanical integrated means of coupling the transfer and joining station operation and timing to the index-dwell cycle of the machine.
- Some provision for mechanical integration of the inspection functions to the index function on the machine.

Figure 3-21 illustrates a typical integrated linear assembly machine. Its modular construction permits the machine to be easily configured to the job requirements.

Actuation of Individual Stations

The mechanically integral assembly machine coordinates the index-dwell cycle of the machine to actuation of individual work stations by means of some mechanical coupling such as a camshaft, sprocket, or gear (Figure 3-22). This mechanical integration eliminates the need for sequential control systems and expensive interlocking systems. It also means that the index portion of the machine cycle can be used by transfer stations to obtain new parts from the feeding mechanism for the next machine cycle, and position these parts for insertion prior to the dwell portion of the machine cycle. This ability to utilize the constant camshaft rotation of integral machines to prestage components for transfer significantly enhances production capability of such equipment.

The ability to have transfer stations reach out to the part feeder escapements for new parts during the index portion of the machine cycle presupposes continuous cam-shaft rotation throughout the entire machine cycle. This is a critical consideration. Since most intermittent motion devices have mechanically fixed index-dwell ratios, the portion of cycle devoted to indexing is proportional to the time required for the work done during the dwell portion of the cycle. If this dwell period is relatively long, the index portion will also be proportionately longer, often at rates far below the maximum possible index speeds imposed by machine design and inertial consideration.

Some builders have chosen to reduce the portion of the machine cycle time devoted to index by placing a clutch between the drive motor and intermittent motion device. The machine is indexed at the highest practical speed (shortest index time) and then the clutch disengages the motor from the machine for whatever time period is required to do the work. This gives optimum index-dwell ratios, but does not provide the necessary continuous camshaft rotation necessary to integrate station operations with the dwell period of the machine. The machine becomes downgraded to an externally sequenced machine. Others have utilized two speed drive motors, but the time required to go from high speed to low speed and back prohibits this approach in all but the slowest systems.

Compatibility with Control Options

Each manufacturer of integral assembly machinery has tried to build into his or her line of machinery and accessory stations motions to ac-

Fig. 3-22. Drive sprockets with adjustable *verniers* are used to couple camshaft rotation with the intermittent motion of the indexing mechanism.

commodate the various transfer, joining, and inspection requirements commonly found in mechanized assembly. For the most part, these motions are cam controlled and mechanically integrated to the index-dwell function of the machine. Such machines require little of the time sequencing control necessary in fluid-power actuated or special purpose machinery. Transfer units move in and out and inspection probes rise and fall in mechanical synchronization with the workpiece transfer device on the machine.

When it comes to tooling such integrated machinery, however, it is extremely difficult to avoid the use of fluid power electro-magnetic operated devices to actuate pick-up fingers and operate escapement devices. Additionally, press stations may have to be locked out and inspection probes activated only when in contact with a workpiece, not just where the workpiece should be. It is conceivable that much of this might be achieved by ingenious mechanical design. In practice, however, it is more inexpensive to couple the assembly machine to a control system with either pneumatic, hydraulic, electrical, and electronic components, or a combination of these elements. The nature of possible control options is discussed in Chapter 4. It is important, however, that in selecting an integrated machine system to determine means by which the proposed machine can be coupled to the selected control system. Rotary cam switches (Figure 3-23) were used extensively to couple camshafts with control elements. The widespread use today of microprocessor based programmable logic controllers has led to greater utilization of magnetic or optical shaft encoders that report main camshaft position to the machine control system.

Cam Design

Integrated assembly machine systems use cam actuation to integrate tooling stations with the machine index system. In some center column machines, single or double barrel cams are used to cause upward and downward movement of the tooling platforms. The throw of this cam will determine the maximum rise and fall of this platform (Figure 3-24).

In practice, a cam will be designed so that the operation requiring the largest vertical travel plus necessary tooling clearance will determine the cam path. Lift and index are considered necessary evils while drop and dwell (or feed) are functionally useful. Unfortunately, the amount of cam degrees available for dwell, where most of the work is done, is what is left over after lift, index and drop are determined.

The smaller the diameter of such a cam, the less the degrees of rotation that will be left over for dwell. Lift (pressure) angles cannot exceed 45° or the cam roller will be placed in shear. It is important to remember that *for a given lift, the larger in diameter or periphery the cam, the less the degrees of camshaft rotation required to produce that lift*. Other operations on the machine with one or two main cams requiring less lift or horizontal motion, will have to have some lost motion, overtravel or compensation built into the station design. *More simply, the larger the cam diameter, the greater the percentage of total cycle time available for fixture dwell and assembly and test activities.*

While dial machines are limited to a very few available cam paths, linear machines can have upper and lower camshafts, each with a large number of double-faced cams to control horizontal and vertical motions. This machine in Figure 3-25 can have up to six double-faced cams working at one time on any given station.

It is important to repeat, the larger the cam diameter, the more degrees of cam path that are available for dwell (or feed). A machine with large cams may be able (because of longer dwell time available) to overcome a line balancing problem not possible for a machine with smaller cams operating at the same cyclic rate.

FACTORS CONCERNING SYSTEMS CHOICE

If we assume that the machine user has a choice of machine systems, there are several factors that should determine his or her ultimate choice. It may be that there is little or no choice for the customer in the system selection. If policy, delivery, price, or experience indicates that the machine is to be built by a specific builder, the chassis selection may become automatic. Increasingly, surplus machine chassis available in corporate surplus machine listings may have

Fig. 3-23. Rotary cam switches were commonly used to correlate control functions with main camshaft rotation. Today's microprocessor controls usually use information from a shaft encoder or resolver.

to be used for the next available project, whether or not they are the best suitable system. Corporate standards or user standardization programs may preclude the user or builder from using a newer type of chassis than that currently in use.

All of these possible restrictions notwithstanding, there are six major areas that should be part of any selection process for assembly systems.

Line Balancing

On any multiple station machine whether it be metal cutting, assembly, inspection or a combination of these, the single, longest operation will dictate machine cyclic speed, with all other operations being performed at less than optimal speed.

Users or potential users often fail to realize what a relatively small percentage of any total machine cycle is available for actual work on the product. A typical case might illustrate this point.

A small dial-type assembly machine is designed to run at 60 strokes per minute. It has an index function requiring 120° of main camshaft rotation, a rise of $2^1/_4$ in. on the tooling platform requires 94° on the rise portion of the cam, the drop on the cam takes 65°. Only 81° of the 360° camshaft rotation is available for part insertion and joining operations. Of the total cycle time of 1 second, only 22% is available for dwell and work on the product. This is not an exaggerated example.

While most assembly functions are for all practical purposes instantaneous operations such as part insertion and staking, other functions require extended time periods. Transfer functions that require rotation of a part to a specific radial location or that require a part to be picked up in a nest, rotated, and replaced require small but significant increments of time. Joining operations such as resistance or ultrasonic welding, spin riveting, and inspection operations that require specific times or establishment of pressure,

Fig. 3-24. A typical barrel cam. Each cam has at least four major sections—index, rapid drop, dwell (or feed), and lift. Intermediate dwells often occur in the lift or drop portions of the cam.

vacuum, or temperature may require significantly long time periods.

Functionally, index, rise and fall must be determined first so that the available degrees of dwell can be established. The time for the single, longest operation must be determined very closely. The time for the dwell portion of the cycle must be adjusted to this requirement by slowing down the main camshaft so that the time required for the traverse of the dwell portion of the tooling cam coincides with the time requirement of the longest operation. For instance, if the described machine had an operation requiring an available time element of 0.56 second, the fastest cyclic rate the machine can operate would be established as follows:

$$\frac{0.56 \text{ second}}{81° \text{ camshaft rotation}} = \frac{X}{360° \text{ camshaft rotation}}$$

X = 2.48 seconds maximum cyclic rate or 24.19 strokes per minute

This type of calculation must be made for any integrated assembly machine using cams for tooling actuation.

If 24 strokes per minute is not sufficient production, then the possibility of line balancing by some means is required, which might include:

- Multiple machines. In extreme high-production conditions, there are many advantages to having multiple machine lines.
- Multiple tooling. Producing two or more assemblies per stroke on the same machine.
- Selecting a chassis requiring less degrees of camshaft rotation (because of larger cams) for index (90° instead of 120°) and large diameter tooling cams requiring less degrees of rotation for rise index, would mean 182° were available for dwell. This would mean a machine capable of operating at 56 strokes rather than 24 strokes, a dramatic improvement.
- Determining the longest operation could be done in two or more steps. For instance, if it were spin riveting requiring 0.56 seconds, could it be done in over two stations of 0.26 seconds duration each?

Each of the preceding examples is for very short cyclic periods and high operating speeds. When we talk of longer cyclic rates and lower gross production rates, the consideration swings to the other side. At high speeds, the short time available for work on the product is limiting. When lower production rates and longer work periods are indicated, productivity is inhibited because of the extremely long index periods on machines with integrated camshafts. The machine will be indexing below its maximum capability. If a very long time period is involved for a specific operation, one which is inconsistent with high overall production rates, line balancing can be accomplished by one or more of the following:

- Using parallel tracks or a series of multiple tooling stations on a power-and-free chassis.
- Ejecting the subassembly to a continuous motion turret type station from whence it is returned to the main machine line.
- Stopping or decelerating the main drive motor during the dwell period for the time period required, but having the machine index at the highest practical speed.
- Ejecting subassemblies to an off-line operation and then returning them to the main machine system.

Fig. 3-25. Adjacent cams are used to place and remove an expanding mandrel on a brake piston to facilitate "O" ring or chevron washer insertion.

It is impossible to ignore the correlation of system choice, overall production requirements, and time periods for the longest operation in the assembly and test sequence.

Part Size and Weight

The size of the assembly and its weight when combined with the size and weight of the necessary work-holding fixtures will often immediately limit the choice of machine system.

- Dial-type or rotary machines can handle parts ranging from the very smallest to those that are moderately large.
- Synchronous in-line machines with cantilevered fixtures cannot handle heavy parts without some form of outboard fixture support. These supports become somewhat self-defeating, since they add additional weight, create maintenance and operator safety problems, and often indicate that the machine is being stretched beyond its designed capacity.
- Over-and-under machines can support quite heavy work loads up to the point where the weight of the whole system causes inertial problems at the start and stop of the index.
- Walking beam machines, while normally applied to moderately small parts, are capable of fairly heavy loads.
- Transfer or power-and-free machines can handle quite heavy work loads.

Volume Requirements

It is a happy coincidence that most assemblies made in high volumes are small in size. Larger parts are usually made in a wide variety of models, each offering different options and features and, hence, the production requirements for any model or family of models is low or may be done in several plant locations. One ball point pen manufacturer may make several hundred million identical products in one year at one plant location. Disposable medical syringes might fall into this pattern. On the other hand, while the total number of cars made might be in the low millions, they are spread over several manufacturers, hundreds of different models and options, and assembled at dozens of plant locations.

The higher the production, the greater the incentive to go to synchronous, mechanically controlled machines. The larger the part becomes, and the lower the production rate is, the quicker economic justification balance will swing to nonsynchronous machinery.

Integration of Manual Operations

While many new products are specifically designed so that all component parts can be handled automatically, there will always be a significant number of products having components that may require manual handling.

Safety considerations and production requirements often preclude the operator from loading directly into the machine fixture nest. Any choice of machine system must include examination of the ability of that system to ergonomically couple the operator with the machine through the use of conveyors, rotary-assist tables (Figure 3-26), or magazine tracks. A part that would be difficult to load into the fixture at 20 or 30 strokes a minute might be hand fed into a conveyor at 40 to 60 pieces per minute whence it can be transferred automatically to the machine work-holding fixtures.

Product Life

It has been the author's experience that industry is overly pessimistic about product life cycles. *Most products tend to have product life far beyond original expectations, and most engineers constantly underestimate the difficulty and time frame for putting a new product into production.*

A realistic evaluation of the product life cycle can be a most significant factor in choosing the assembly system. Short-lived products might well be handled best by good manual assembly lines, assisted by inexpensive conveyors and sound industrial engineering. Dial feeds on single-station joining machines, such as presses, screwdrivers, and welding machines are the next logical step. Small fluid-power-actuated machines are the next possible consideration.

In each case, minimal capital cost is balanced against high assembly labor content on the assumption that product life is too short to obtain sufficient capital recovery from the purchase of a major assembly machine or system. If the prod-

Fig. 3-26. A rotary assist dial allows an operator to radially orient bomb fins before insertion into the main machine. Simplified fixtures aid in orientation and also provide float for the operator in keeping pace with the machine.

uct life proves longer than anticipated, and particularly where a competitor has decided to mechanize assembly, the choice of manual assembly is poor indeed.

Product Design

It has been mentioned earlier that it is essential to have some degree of product design stability if there is to be sufficient lead time for machine development and capital recovery. It is becoming increasingly common for products to have a continuing evolution often accomplished by small changes in existing components or addition of optional components to meet the changing needs of the buying public and match new features offered by the competition.

In selecting a machine system, it is necessary to consider the ability, time, and cost of modifying the machine in the field, on the operating floor, to accept running product changes. It is worthwhile to repeat that tool engineers, and particularly tool designers, tend to solve the problem at hand. The customer must be certain that the chosen system has the capability of accepting running changes in product design and also be capable of accepting totally new products, should a new product do to an existing product what small electronic calculators did to the slide rule business.

Salvage value of assembly systems is another consideration, but the potential user must be aware that the major cost of any significant assembly machine is not in the chassis or controls, but in the parts of the system unique to the product. Work-holding fixtures, escapements, pick-up fingers, feeders and gauging equipment, joining stations design and debugging, all make up the major cost of any assembly machine with very modest salvage value.

In the case of a major product redesign mak-ing an existing assembly machine obsolete, a standardized machine may be readily converted to the new product. This may be as high as 90 or 95% of the former standard machine and tooling usable in the new product. While the physical standard components of the machine reutilized may exceed 75%, rarely is there more than 30–40% of the original purchase cost (stated in current dollars) usable on new products.

SUMMARY

The machine chassis selected is the foundation on which the control system, the work-holding fixtures, and the tooling stations are seated. It must have those properties that simplify the control requirements, accurately locate the fixtures, and facilitate the coordination of tooling stations. It must, in itself, be an extremely dependable element, not adding to the operational problems that are a day to day fact of mechanized assembly. The chassis should be energy efficient, quiet, and capable of future modification and growth. The user will face a bewildering variety of commercial offerings, which must be evaluated in the light of the guidelines given.

In the increasing use of automated assembly systems for complex assemblies, it is not at all unusual to find rotary and linear systems, synchronous and power-and-free systems, multiple tooled palletizers and depalletizers tied together in a single, large integrated system with buffer storage type conveyor systems. As assemblies become more complex, the experienced builder must have a complete kit of system approaches.

REFERENCES

1. K. Treer, "Selecting Intermittent Transfer Devices" *Automation* 61–67 (December 1962).

Chapter 4
The Role of the Control System

INTRODUCTION

The electrical controls in an assembly system play a significantly different role than the control systems in other types of machine tools. They not only control the stopping, starting and sequencing of assembly machine operations as in other types of industrial equipment but the control system also monitors in a unique way the quality and functionality of the product being assembled, causing some form of corrective action when the product fails to meet assembly or functional specifications.

To a significant degree, the choice of control design will be dictated by the chosen machine system itself and the sequence of the assembly process. Cam-actuated integrated assembly machinery will require little control sophistication for basic assembly functions, for machine index and transfer motions. Machines based on power-and-free transport systems and independently actuated work stations will require a more extensive but relatively simple control system to integrate machine functions.

A most significant and critical assembly system control decision will reflect the customer's and builder's view of probable assembly component parts quality levels, their view of probable system efficiency, realistic assessments of probable net production levels and the cost or practicality of defective product salvage, correction or repair. Initial correct design choices in this area which provide solutions to assembly failures will significantly reflect the experience level of the machine builder and user.

One often overlooked major factor determining ultimate machine production is a basic premise:

*The purpose of assembly machine controls is to **operate the machine.***

In order to do this the controls must be so designed as to *assist the operator in operating the machine.* Control operating stations should be designed with the following features:

- Ease of control-operator interface to the point that control layout provides almost intuitive instruction in operating the equipment,
- Control panels should be able to pinpoint the precise problem, whether operator curable or technician dependent, in the shortest possible time.
- Ease of machine set up and production model changeover.
- Ease of trouble shooting for any electrical malfunction.
- Ease of calibration and verification of gauging and testing stations.
- Fail safe testing, inspection and gauging stations.

Control selection is one of major importance. In the days of hard-wired control panels, a design choice made in error usually was permanent. Today's proper use of expandable microprocessor based control systems provide for continuous system improvement. *Control choice is, together with tool design, the basic determinant of net production levels realized to the degree that it makes provision for assembly failure as a normal component of machine operation.*

Before any rational choice of controls can be made, several fundamental decisions must be reached. The first is to select a specific assembly system—the means by which the assembly is moved from operation to operation. Basic types of assembly machines discussed in Chapter 3 require different control approaches.

An examination of assembly size and weight, the feasibility of automatically feeding component parts, joining or bonding times, volume requirements, and potential savings will indicate which type of basic assembly machine design is most suitable for a given job. A short-lived product manufactured at moderate production rates might indicate a simple semiautomatic fluid power approach. High-volume production requirements, combined with a lengthy production run, will probably indicate the more durable mechanically actuated, synchronous-type machine. Severe line imbalance or heavy parts, which may indicate integration of manual operations or off-line repair loops, will point to some form of power and free palletized system.

As mentioned in Chapter 3, rotary (dial-type) and fixed pallet, in-line indexing machines may or may not be mechanically integral in their operation. They are integral if the motion of the various transfer devices, inspection probes, press stations, and joining devices are actuated by the same power source that controls indexing of the work-holding fixtures. All movements on such machines are synchronized by common camshafts or gearing. (Figure 4-1.)

Other rotary or fixed pallet machines have work stations individually actuated by solenoids, motors, cylinders, or other electromechanical or fluid-power devices. These stations are actuated on signals from the intermittent motion device that indexes the fixtures. Such machines as the one in Fig. 4-2 require larger and more complex controls than mechanically actuated machinery.

Free pallet machines, often called power-and-free, such as the one shown in Fig. 4-3 are generally used in two situations. The more important situation would be to address line balancing requirements, where various assembly and joining operations differ widely from one another in time requirements. In power-and-free systems different techniques are used to balance long and short operations. Multiple transfer, joining, or test stations can be banked, or pallets sent down parallel tracks for lengthy operations while shorter operations are done on a one-at-a-time basis. Secondly, since there are less restrictions on product weight in power-and-free design size, these machines are used for modest volumes of heavier assembly operations. Free pallet machines are usually non-synchronous (a station operates only when a pallet is positioned before it), have lower cyclic rates than synchronous machines, and are often semiautomatic, combining manual and automatic operations.

Special purpose assembly machines include single-fixture machines performing several operations, multiple insertion machines, and continuous motion machines. It is difficult to give specific suggestions regarding control systems for any assembly system before mechanical designs have been established and a specific machine design chosen. These may range from the smallest microprocessor based controller to large multi-processor distributive systems.

The next logical step before selecting a machine control application is to analyze the scope of operations required of the control system. A requirement for all machines is to initiate, coordinate, and stop the various motions of the machine and its transfer and joining stations. Secondly, various control elements will be added to ensure that the machine does not continue to operate in an unsafe condition. Thirdly, the machine will be required to monitor and identify incomplete or defective assemblies (see Chapter 8) in a way that provides the highest net production of acceptable assemblies in the most reliable way, and at the lowest capital and operating cost.

The design and construction practices for such control systems as well as safety guarding must

Fig. 4-1. A mechanically integral assembly system where all indexing, transfer and inspection motions are controlled by a single power source.

be done in conformance with local (or national) standards, meet all OSHA requirements for operator safety and conform to the NFPA national electrical code. It is in the area of determining product quality (the inspection for presence and position of component parts, or end-item functional, dimensional, and test characteristics) at each incremental step of added value, that assembly system controls are unique. It is not only a verification role at each incremental step of assembly but increasingly an active corrective role for identified defective assemblies.

The system builder's in-house ability to integrate a large number of computer controlled transfer, joining and test stations through integration software is a true mark of the builder's control design competency. Creation of such integration software

for the products of many vendors without any commonly accepted manufacturing software protocol is the most demanding engineering task for today's assembly systems design.

In preparing this chapter for the second edition, lengthy discussions were held with colleagues regarding the structure of this chapter.

One suggested approach was to address assembly system controls from a product design and best practices approach. Such an approach would need to address such functional issues as:

• Basic controls
• Integration of stations
• Inspection of processes, components and product function

Fig. 4-2. Large transfer lines require control systems to perform sequential actions, not necessary in a synchronous machine.

- Data acquisition and statistical analysis
- MIS reporting
- Supervisory control of large integrated systems.

It would also mean a state-of-the-art report in current control technology including:

- PLC's
- Microprocessor based controllers
- Motion control
- Sensor technology
- Architecture (distributed control)
- Vision systems
- Simulation and expert system.

Additionally, it would have to address system design standards, object oriented (modular) controls, machine memory systems, indication and documentation of process performance and production and cell control issues.

The scope of such a comprehensive review if properly done would probably represent a major book in itself and face the danger of immediate rapid obsolescence in the ever changing world of manufacturing controls hardware, gauging and testing systems and operating software.

For two reasons this chapter instead continues its original focus on underlying principles for control design selection. One is the reluctance of many assembly system customers to face the real probabilities of component insertion and product function failure during the assembly process. The second is that relatively few engineers on system acquisition teams are highly conversant with the range of state-of-the-art commercial control systems and rapidly developing trends in gauging, testing and manufacturing software.

This chapter, therefore, is intended to address the consequences of insertion and product function failure in assembly systems and the control

Fig. 4-3. This engine head assembly system, as all free pallet machines, requires extensive interlocking as well as sequential control.

approaches that will result in the highest levels of net production of acceptable products.

INSTANTANEOUS VERSUS MEMORY CONTROL SYSTEMS

What should the machine do if it fails to feed or improperly positions a part? If the inserted component does not match the receiver component? If the completed assembly fails to meet dimensional or functional specifications? If a component part is missed, should the machine stop, or should it lock out subsequent operations, until ejection or should it send the incomplete assembly down a separate path for rework while continuing to work on other assemblies?

Instantaneous Control Systems

Stopping the machine immediately upon detection of any malfunction in the assembly process is easily done by opening the holding circuit on the run button. If the inspection device indicates an improper assembly condition, indicator lights, message displays, operator interface terminals and man-machine interface panels or terminals properly labeled identify the causes of stoppage. The most simple system, reliable in operation, low in cost and simple to maintain, is called an instantaneous or 'stop and correct' control system.

The primary argument against the use of simple stop and correct systems claims that low efficiencies in feeding or joining stations lead to excessive machine stoppage and lower production capabilities. This claim is indeed serious if proven. For example, if a machine costs $240,000, with a projected 14 months return on investment when operating on a two-shift basis, it is expected to return some $51 an hour in savings. If an improper control system adds only two additional minutes of downtime per hour, projected returns will fall short by some $8000 in that period. This shortfall will continue throughout the production life of the machine.

Memory or Logic Systems

Those who argue against immediately stopping a machine at the first station failure favor some sort of memory or logic system. The argument is that if a machine fails to insert a part, thus causing the machine to stop, not only is that assembly lost to production, but the good parts that could have been produced if the machine was kept in operation are also lost. Memory systems are intended to keep the machine running should a defect occur, while segregating the defective parts at ejection. Theoretically, then, only defective cycles are lost.

The unstated premise for this pro-memory argument is that most station failures will be random and self-clearing. In other words, if a station fails to feed or join properly on one cycle, it will clear itself and function properly at the next cycle. Unfortunately, in making controls decisions, many overlook this implicit premise.

The basic judgment is this: Will this promise of self-clearing failures hold true in fact? In deciding which of the two approaches is the proper one, the manufacturing engineer must ask: Will most assembly failures be self-clearing and self-correcting? If they are not, how many defective machine cycles should occur before stopping the assembly system for corrective action?

It is not so much a choice between instantaneous control systems versus memory systems, but rather what control scheme you want to use on any station if any. All machines must have the ability to stop immediately upon sensing an improper or unsafe operating condition. Air pressure failures, improperly opened guards, incorrect location or relationship of machine elements or tooling, and failure of electrical interlocks to open or close at the proper time would be typical conditions demanding instant stoppage.

Machine Memory. Memory systems add several functional elements to the control system. Machine memory used to identify or "remember" where any individual moving work-holding fixture is, relative to the machine's fixed operating stations. Machine memory tracks a fixture and condition of the assembly it contains as the fixture moves through the machine. Information sent into the control system by various monitoring inspection devices is processed, and then assigned to a specific fixture.

A signal (or lack of signal) may indicate, for example, that the developing assembly in work-holding fixture 26 did not correctly receive a component part at feeding station 3. Depending on the logic programmed into the system, one of several alternative actions are taken, based on information received from the inputs and shift register. Continuing the same example, when fixture 26 arrives at feeding station 4, an output module, accepting the information that fixture 26 has an incomplete assembly, could lock out feeding station 4 and all subsequent operating stations. Outputs may be used to lock out subsequent feeding and joining stations, actuate alternative ejection units, color code or otherwise identify parts by classification or defect, or let parts go around for a second assembly attempt. For instance, the incomplete assembly in fixture 26 may be carried back to station 3 for a second try.

Consecutive Failures. Any validity to the choice of a memory system must be at least based on the assumption that a significant portion of feeding problems in parts transfer stations will be random, not consecutive, failures; specifically not those which indicate a jam in feeders or tracks, inoperative or misadjusted transfer or joining units, or worn and broken machine parts. When insertion failures are consecutive, some corrective action is required before the machine can operate properly. Consecutive malfunction detectors shut down the machine after any programmed number of consecutive malfunctions. Rarely are more than three consecutive misses accepted before system shut down. *Any indica-*

Fig. 4-4. A mechanical transfer unit can be locked out by using an air cylinder instead of a mechanical linkage to restrict vertical travel, or by locking out the air actuated pick up fingers.

tion of inspection station failure must stop the system instantly. All inspection stations, gauging and functional testing stations must be self-inspecting at each cycle of the system (usually during fixture index). They must be integrated in the overall system design so their failure would indicate a bad product. An important element of any good memory system are counters to record total cycle counts, total acceptable or defective assemblies, and the number of failures at each working station. Such counters are invaluable for troubleshooting purposes.

Station Modification for Lockout Purposes. When all station movements and/or functions on an assembly machine are controlled through fluid power or electrical means, locking out downstream stations after detecting a malfunction on a memory-controlled machine is quite simple. It is merely a change in the functional signal for those specific stations.

In machines that are essentially mechanical and cam operated, the control interface becomes somewhat different. A totally mechanical station integrally connected with the machine chassis index system cannot readily be locked out unless some component of its construction, some linkage between gearshafts or camshafts and station work face, can be modified by electromechanical or electropneumatic means. Figure 4-4 illustrates two areas by which a transfer station can be locked out. In one approach, vertical travel is inhibited by a cylinder linkage, and in the other, normal transfer motions are made nonfunctional by failing to close the pick-up fingers at the normal time.

Mechanical press stations are somewhat more difficult to lock out, but one way is to remove the toggle or pivot point of the press. While the ram will continue to move, it is not capable of exerting any force. Figure 4-5 illustrates one such lockout system.

Alternate Ejection of Defective Assemblies. Whenever the chosen control system is totally instantaneous, that is, it stops on the detection of

Fig. 4-5. A pancake cylinder is used to lock out this mechanical toggle press when an incomplete assembly is present.

any malfunction, whether it be insertion, joining, or functional testing, it was usually assumed that the operator would correct the parts either in the machine, by substitution of another assembly that is correct, or by discarding the defective assembly. There was no need for alternate ejection. Today's powerful microprocessor based controllers offer a variety of alternative actions.

The need for providing an alternative ejector first occurs when the control choice is to stop the machine after assembly defects are discovered but to allow the machine to continue to run upon detecting a nonfunctional or nonacceptable as-

sembly, by segregating ejected assemblies into acceptable and nonacceptable containers.

It was considered good industry practice to place alternate ejection stations prior to good parts ejection. The reasoning was that it was better to eject a good part erroneously than to allow a defect to be ejected as good. That sequence has been changed by the increased number of control options available on detection of assembly malfunction, or product function defect.

When the memory circuits are used to lockout downstream feeding stations after an assembly failure is detected, then it must be assumed that

Fig. 4-6. Two ejection stations on this machine segregate defective and acceptable assemblies using data in the control system's shift register.

many of the assemblies to be alternately ejected are incomplete. They may be incomplete at any stage of assembly and hence will require alternate ejection in any stage of incomplete assembly. The alternate ejection may often require more complex tooling development than the good parts ejector. (Figure 4-6.)

The stress and growing utilization of matched assemblies and the high value of product components in such assemblies dictate alternatives to simple alternative ejection of incomplete or defective assemblies. The realization of gauge drift or improper coupling of sensors when measuring microscopic differences makes on line repair an increasingly attractive option for selective or matched assemblies. It is also utilized when a critical component may not always be available completely within design control limits.

In such instances, customers may opt for control designs which provide for separate ejection of defective or salvageable, or assemblies for machine repairable defects to be run through the as-

sembly system for a second attempt at correct assembly.

Some of the current ejection options include:

- Segregation of rejected parts into scrap, rework or off-line salvage before final testing.
- Ejection of partially or fully assembled products for off-line audits. These audits may be triggered by elapsed time, specific production levels (i.e., every thousandth part) or on specific demand when done by control panels such as color touch screens to allow ejection off-line for audit at any incremental stage of assembly or testing. Such ejection is done on a full operation run of the equipment, without stopping or system shut down.
- Rework options. Instead of leaving the system, incomplete or defective products are recycled through the system for a second attempt at successful assembly. This may involve partial disassembly, particularly of matched assemblies. Such attempted repair is justified when gaug-

ing technology units cannot determine assembled dimensions on the upper or lower limits of acceptable dimensions or where ambient changes may affect sensors, transducers or control settings.

CONTROL IMPACT ON MACHINE EFFICIENCY AND NET PRODUCTION

Before proceeding in the examination of possible control choices, it is necessary to clearly define three terms:

Gross Production

Net Production

Machine Efficiency

Improper use of these terms leads to a great deal of misunderstanding— probably more misunderstanding than any other terms used in assembly system procurement, with the possible exception of the word "delivery."

If it seems out of place in a chapter on machine controls to be defining these terms, it must be remembered that what is unique to assembly system controls is not the coordination of machine movements, but the machine monitoring of the developing quality of the product being assembled. Fundamental control design choices are based on what the machine designer feels are the most productive ways to handle the information received from the quality inspection probes and sensors.

Let us examine these terms before going to the control implications.

Gross production is an expression of the maximum possible production of an assembly system expressed in units of production in a given time period. The implicit assumption is that each machine cycle is theoretically productive; that is, each cycle is capable of producing a functional, acceptable product. Any assembly system, therefore, that is running at a cyclic rate of 40 strokes per minute, has a gross production rate of 40 assemblies per minute, 2400 assemblies per hour.

This concept seems so simple that it is surprising how much grief is caused by its misuse. A customer may request a gross rate of production. A builder may quote a possible range of gross production rates (i.e., 35–40 assemblies per

minute gross), but the machine's actual gross rate will be the maximum number of cycles it can complete in a given time period without stopping for machine system or assembly defect. This number is clear cut on a synchronous machine with continuous cycling, but may be indeterminate when individual cycles are operator initiated, or occur on completion of the longest operation in a power-and-free system. Pneumatically operated machines that are sequentially interlocked may vary in cycle speed over a considerable range of cyclic rates.

In a mechanically actuated machine the gross speed of the machine may not be its theoretical maximum design speed, but an empirically selected speed determined by motor speeds and pulley sizes or other exterior constraints. That cyclic rate is the actual gross production of the machine for a specific time period. *Any discussion of net production must be discussed as proportional to that gross speed.*

While it is easy to determine what the gross rate of a machine in a minute or in an hour is, it becomes much more difficult to ascertain what is meant by a shift. Is it eight hours or seven and one half? It is 480 minutes or 400 or less? How many working days are there in the customer's week, and how many days in the customer's year?

In brief, the term "gross production" is understood by all, but its specific application may be cloudy.

Net production is the actual achieved acceptable product production stated in terms of a percentage of gross production capability for a specific time period. Simple enough at first glance. In a synchronous machine with no functional testing, it is the number of assembled parts that are ejected stated as a percentage of the maximum number of parts possible if the machine ran at its cyclic rate for that period without interruption.

Once functional testing is included, a very different problem emerges. The user tends to look at net production as the number of good parts ejected, but some builders may feel that net production is the total number of parts ejected, acceptable and defective. The builder will feel, with significant justification, that the machine has done its job, not only in assembly but testing as well. The builder feels that the rejected parts were assembled correctly and hence this loss is a product problem not an assembly problem. One can legitimately use either figure—good parts or total parts—to state net production.

The problem becomes more complex once one includes any configuration of memory control on the machine. One could state net production as the number of completed and acceptable parts ejected, but one would have great difficulty in including the incomplete assemblies as part of net production. The fact remains, however, that a significant number of these incomplete assemblies were created because a nonclearing jam or blockage existed and the machine was programmed to run one or two additional cycles before shutting down, significantly increasing the number of defective parts created.

One often hears the use of "efficiency" as a substitute for "net production." In fact, *efficiency is the statement in percentage terms of successful functionality of any given machine actuation.*

Net production is the result of the loss of gross production capability because of inefficient machine actions and the resultant downtime that is caused by such inefficiencies.

There is no more important sentence in this book than the preceding sentence, and an understanding of what it signifies will underlie any sound judgment when critiquing or selecting assembly machine concepts and design. One of the most significant problems facing designers of automatic assembly control systems is the enormous exposure to assembly failure. Failure of the assembly process can result from overall system failure; station failure; lack of aesthetic, functional, and dimensional characteristics of the parts being inserted; and the inability of the part or the subassembly in the machine fixtures to receive additional components. All of this, in addition to a recognition that components which have been successfully inserted may produce a functionally defective product.

Any of these failures will result in reducing system efficiency and have a dramatic impact on net production.

It is first necessary to determine **machine efficiency** in terms of the percentage of machine cycles that will produce acceptable assemblies.

When C is the percentage of acceptable parts in each lot of component parts coming to the machine, S is the efficiency level of each work station in performing its own task of selection, transfer, or joining, and M is the efficiency of the basic machine control system in coordinating all of the individual station operations, the probable percentage of machine cycles that will produce acceptable assemblies can be expressed:

$$\text{Machine efficiency} = (C_1C_2C_3 \ldots C_n) \times (S_1S_2S_3 \ldots S_n)M$$

It is possible that on a short run basis, or if the quality of components or station efficiency is quite low, a particular machine cycle can be deficient for more than one reason, bringing up the overall efficiency level. However, on a long run basis the probable efficiency level indicated by the preceding formula is quite realistic.

A machine running at 60 strokes per minute assembling 10 components with two joining operations using 20 solenoid valves in the tooling stations might have 3720 areas of readily identifiable possible system failures each minute of operation. It is readily apparent that extremely high efficiency levels in each functional area are necessary for successful machine operation. Excessive emphasis, however, on seeking unattainable efficiency goals tends to overlook the importance of the length of downtime in determining net production on automatic assembly machinery.

The formula for net production given below is oversimplified in order to emphasize that *length of downtime is probably the most easily controlled factor in determining net production.* The number of inefficient machine cycles (C_d) in a given period obtained from determining machine efficiency levels as explained above is multiplied by the average downtime (T_d) that each malfunction causes. The result is subtracted from the total time (T_t) in that period, and the result is divided by gross cyclic rate (R_g).

$$\text{Net Production} = \frac{T_t - (C_dT_d)}{R_g}$$

Average length of downtime is substantially determined by overall system mechanical and control design.

Early in the development of assembly machines those who were aware of the significance of downtime length promoted so-called memory systems as being capable of equating average downtime to the length of one machine cycle; in effect equating machine efficiency and net production. In real life, such assumptions presuppose that failures to insert a component properly are always self-correcting at the next stroke of the machine and that salvage of incomplete assemblies is not necessary or is not costly. The cases in which this can be safely assumed are extremely limited. In practical terms they are non-existent. Lastly, *it cannot be assumed that successful assembly from a mechanical viewpoint alone ensures the functionality of any product.* Product function may consist of electrical or mechanical characteristics, or characteristics determined by such factors such as resistance, hardness, pressure, torque, capacitance, or durometer readings not directly or solely related to physical dimensions. An assembly machine may be required to determine the existence of such characteristics to within design limits, to determine whether the assembled product is functionally useful. In doing this, the machine is performing its designed role of assembling components, but a user may be reluctant to accept that successfully assembled products, segregated because of functional defects, are part of net production.

Some readers will not grasp these concepts at first reading. A solid grasp of these concepts is so important that it is worth restating them in a different fashion.

Assume for discussion that an assembly machine is running at a gross cyclic rate of 60 strokes per minute and that the user has scheduled 16 full hours of production for that day. The machine has, therefore, a gross capability of 57,600 assemblies in that 16-hour period. Let us also assume that this product was composed of 10 components requiring two joining operations and ejection. If we ignore mechanical or electrical problems with gauging stations, fixtures and other ancillary features on the machine, we have 13 identifiable assembly functions each second or 748,800 possibilities every two shifts for something to fail or be *inefficient.* Let us assume that at the end of 4 hours of absolutely efficient op-

eration, a pick-up finger on a transfer device breaks and the machine is down for 4 hours to replace the finger. At the beginning of the ninth hour the machine is restored to operation and runs without failure or stoppage for the next hour. We now have illustrated a machine which ran at a net production rate of 75%, but at an efficiency level of 99.999986%. Only one thing went wrong. *The culprit was downtime.*

Assembly system development must, therefore, not only focus on how to make any assembly system efficient, but also to ensure that system downtime is kept to an absolute minimum. In the hypothetical case above, having a spare set of pick-up fingers available was the solution. In actual practice, however, given sound machine design, and reasonable maintenance, most *inefficient machine cycles will be related to failure to feed parts or insert them properly.* Advocates of memory systems state that they reduce downtime to the length of the inefficient cycle and let the machine self-clear at the next stroke. *This assumption of the predominance of random self-clearing failure must be analyzed.*

The Probability of Random Failure

Random failures, occasional misfeeds or incomplete joining operations, must have some cause. If truly random and self-clearing at the next machine cycle, the identification of these causes and their cure may be more costly than the trouble the failure causes. Usually, however, most failures have easily identified causes: excessive variation in component parts, chips and foreign material, poor hopper construction, misadjusted or worn transfer units, or erratic operation of pneumatic or electro-mechanical actuators. The manufacturing engineer must decide whether to attack these problems directly, or choose an indirect approach, the acceptance of a significant number of assembly failures accommodated by a memory control system. Justifiable reasons for choosing the indirect memory system approach might include ease of defective product salvage, the difficulty and/or excessive cost of improving component part quality, and the willingness in certain industries to accept relatively high scrap levels.

These initial reasons for selecting a memory system are, as mentioned earlier, often based on the premise that after a deficient machine cycle subsequent machine cycles will produce good assemblies. If this is not so, if a solid jam has occurred in a feeder or conveyor or a machine structural element is damaged, return to full production is actually delayed because the machine will not shut down and identify problems until a programmed number of failures has occurred. Assuming the same downtime to clear such jams or repair such damage as in the case of instantaneously (stop to correct) controlled machines, memory-controlled machines conceivably will often have less net production than those that stop instantly.

A more serious problem of random failure is the marginal station in which failure is not merely occasional, but frequent, and in a supposedly erratic pattern. Marginal feed rates on hoppers, worn tooling, marginal PLC timing, worn fixtures, or similar causes result in stations where problems may occur as often as every second, third, or fourth cycle. Net production drops, while the logic system keeps the machine in operation, trying to meet production goals. Since these problems are not resolved without positive corrective action, *a vicious cycle of trying to meet increasing production goals with declining capacity is created.* Instantaneous control systems demand immediate corrective action on problem areas if machines are to run at all.

Experience indicates that on a well-debugged and properly adjusted machine, *relatively few feeding or joining station problems are self-clearing.* Where assemblies are of the additive type (parts placed one on top of the other), more production at lower cost will usually result if efforts are concentrated on sound machine design and consistent part quality, than if efforts are directed toward memory systems. When extensive testing is done after assembly, or when product design and inexpensive salvage costs permit continued assembly (even with an occasional failure), memory systems may provide increased net production.

Impact of Product Design and Salvage Costs on Control Choice

Justification for the type of memory control scheme is dependent on the anticipated probability of significant random failure, but also on the nature of the product design and the feasibility of salvage of incomplete assemblies. There are essentially two types of assemblies (if one is willing to accept a significant number of assemblies incorporating features of both types). These definitions, spelled out in Chapter 2, have dramatic impact on control system choices.

In the first assembly configuration, one part is placed on top of another. If any component is missing, there is little point in continuing the assembly without some corrective action. This action may be manual placement of the missing part or may be a lock out of all subsequent placement operations with alternative ejection of incomplete assemblies. This type of assembly is called an additive assembly, one part being added on top of one or more parts. Figure 4-7 illustrates an additive assembly.

A second type of assembly can be referred to as a multiple insertion assembly, which consists of the insertion of many components on a common base. The most common example is a printed circuit board. Here the success or failure

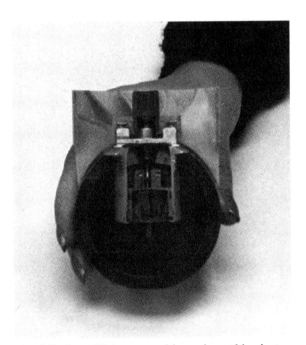

Fig. 4-7. An additive assembly such as this cluster bomb is designed for automatic assembly with each component entered (or added) in the same direction.

of any one component to be inserted has no impact on the insertion of other components. Detection of insertion failure may indicate a lock out of subsequent operations or more commonly, a continuation of the assembly with segregated ejection of incomplete assemblies for salvage or rework. *The choice of control system and programming, therefore, is significantly dependent on the type of assembly and the cost of product salvage.*

Figures 4-8 A and B illustrate multiple insertion assembly. In Fig. 4-8A, eight tubular rivets are used to hold the friction pad of a disc brake, while a ninth is used to retain the sounding plate. Tubular rivets are notoriously poor in quality. With some several hundred being fed each minute, failure to feed a rivet was not uncommon. Failure in one location, however, did not affect other locations. Memory here is practical since missing rivets in incomplete assemblies could be readily added by a bench riveter at low cost. Salvage was technically and economically practical. In more expensive products some form of automatic rework option could be utilized.

Another such instance is shown in Fig. 4-8B. Missing interconnect wires could readily be added in a bench operation. For this reason, sal-

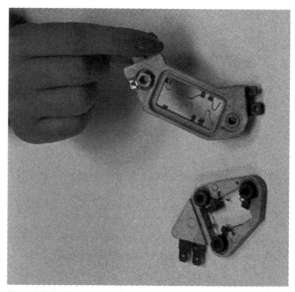

Fig. 4-8B. Terminal housings for a solid state automotive controller has a series of small wires welded to metal terminals prior to the installation of the substrate.

vage of an assembly with one missing wire was economically and technically practical.

In both cited instances the selected system made no attempt to lock out downstream operations as would be necessary in an additive assembly. Memory control here provided high productivity, since random failure was frequent but low salvage costs economically practical.

MACHINE SEQUENCE CONTROL

Since assembly machines are operating at speeds far below computer speeds, and since most control decisions are determined by simple binary inputs, fluidics were widely considered for assembly machine control systems in the late 1950's. Their application was hampered by the various diversity of systems offered commercially, and problems in ensuring clean, dry air for their operation. However, fluid controls had unique capabilities. For example, they still are used to control equipment in hazardous or explosive environments. They are frequently used as flow sensors and non-contacting detectors for parts presence. Since they work on air-pressure differentials, they may be used to sense part pres-

Fig. 4-8A. A disc brake subassembly has eight tubular rivets to retain the friction material to the support plate. Success in feeding any given rivet does not affect the placement of other rivets.

ence, orifices in component parts, leak rates, etc. For instance, deterioration in inertial spin rates of instrument motor rotors can be detected by fluidic probing of small holes in the rotor. Physical contact and photoelectrical probes could not be used here. Fluidics will have a continued if significantly reduced application particularly in explosive environments, but problems in air supply and air quality and present costs limit their general usage. Their use will generally be in specific station applications, rather than for full control systems. Today's control systems are microprocessor based electronic controls. Such controls increase both capability and operational difficulties.

Final debugging of control system problems must be done on the customer's production floor. Initial production runs will expose problems not seen in debugging. This requires, at the time of purchase, a frank, honest judgment concerning the on-site availability of customer's electrically and electronically trained troubleshooters and the availability of necessary test equipment on the production floor. Customer engineers often opt for systems which require highly trained technicians, while the corporation constantly is reducing skilled trades personnel.

Electronic noise suppression may be required, and this often must be done empirically once the equipment is installed on the production floor since full internal shielding may be difficult. Experienced builders find customers are not often sensitive to lack of electrical noise shielding, voltage spikes, brownouts or lack of a strong positive ground system for advanced electronic controls.

Incorporation of dataliners or message displays in the control system specifically to pinpoint the location of specific electric failures, such as proximity switches, solenoids, push button, relay coils and motor overload devices, and PLC input or output card failures eases troubleshooting and reduces system downtime significantly.

From a builder's standpoint, the microprocessor based controller allows excellent flexibility in debugging, since programs can be entered, deleted, or changed in minutes with simple manual input devices (Fig. 4-9). In the field, troubleshooting can often be accomplished by simple

telephone modem links to centrally located computers. Standardization of the programmable controller elements or cards eases replacement and spare parts availability problems. Customers find standardization of control panels through the use of programmable controllers very helpful when electrical maintenance personnel are assigned from maintenance pools. This standardization permits minimal training and stocking of replacements, and can justify higher initial capital costs.

One highly potential opportunity is the realization of the untapped potential of large computer systems or LANs to acquire assembly system production data. Most corporations have large computer installations to handle data processing requirements on a real-time basis.

Microprocessor Based Controllers

In the development of assembly machinery enormous effort goes into analysis on the machine's ability to handle the routine variations in component part quality. Continuous, small but critical, modification to machine elements and operational sequences are normal to the debugging process. There are practical limits to such modification. At some point continued modification of the system and the controls does not result in identifiable productivity increases. The failure rate at this point is now primarily caused by component part quality.

The debugging process in itself places demands on assembly machine control design. The nature of much debugging, often trial and error, means continuous modification of control requirements. The inherent flexibility of microprocessor based controllers is extremely useful, particularly when it can be achieved without physical rewiring.

When in the debugging process machine redesign efforts have reached a point of diminishing returns, there still remains those assembly failures due to part variation. Unless failures to assemble are primarily random and salvage costs proportionately low to salvage value, machines probably should stop on insertion failure for manual corrective action. In the real world, unfortunately, actual determination of these facts can only come after deep analysis of significant production runs. An optimum assembly machine

Fig. 4-9. A representative microprocessor based machine controller capable of machine control, alternative assembly sequences and data reporting.

control system must be adaptable to utilize either instant stop or memory-type controls modes on a station-by-station basis.

This flexibility requirement during debugging must extend throughout the production life of the machine. New molds, changes in vendors, material changes, special production orders, and end-product evolution may change control requirements on a frequent basis. This is a way of life under concurrent engineering programs and a recognition of the importance of continuous improvement activities.

The control system should optimally be adaptable to complete or partial retooling of the basic machine system with a maximum salvage value, hopefully done by sensor input changes and reprogramming.

Another unique requirement for automatic assembly is that of visual displays to record operating characteristics. Specific counts of failure rates at each station are necessary for objective analysis of machine operation. Operators need quick visual identification of what sensor caused the machine stoppage. Lastly, input and output devices should have graphic displays of mode status for control diagnostic purposes.

The manufacturing engineer faced with the choice of a proper assembly machine control system has a wide variety of control options. Before selecting a system, he or she must evaluate what degree of control will result in a proper balance of capital and operating costs with the best net production levels of quality end products. The decision cannot be made without a good grasp of the probable quality levels of incoming parts, the skill and motivation of machine operators and maintenance personnel, and the corporate goals in industrial relations, product quality, and production levels.

ASSEMBLY PROCESS CONTROLS

In nonsynchronous assembly machines, station functions are normally sequentially interlocked. That is, completion of one movement will trigger another movement and so on until the cycle is complete. Index then occurs and another cycle is initiated. Microprocessor based controllers are ideally suited to control these sequences.

In synchronous, mechanically actuated machines, however, machine motions are in the main cam controlled without any direct connection to the machine control system. It is necessary, therefore, to tie in the electricals, electromechanical, and electropneumatic components of the machine to the purely mechanical portions of the machine.

Sequencing of Work Stations. The interfacing of a synchronous machine main camshaft or gearshaft to units that are activated or deactivated electrically is done in one of two ways: by the diminishing use of rotary cam switches, or by optical encoders or magnetic resolvers.

Shaft Encoders. The use of shaft encoders or optical resolvers directly coupled to maintain camshafts or indexes has been found to offer almost limitless sequential control on synchronous assembly machines equipped with programmable controllers. A magnetic resolver is in construction similar to a motor or generator. Stationary field coils establish a plane of reference from which the angular position of the rotor can be monitored. This analog signal can then be converted to digital form for connection to a programmable controller. Figure 4-10 illustrates such a system. Similar input can be done by use of an optical encoder, which counts the passage of graduations on a rotating disc.

The shaft of the encoder or resolver is coupled to the mechanical camshaft of the machine system and thus provides a digital signal indicating

Fig. 4-10. A shaft encoder measures the angular relationship of the rotating shaft (which is driven from the main camshaft of the assembly machine) to fixed stator windings in the housing. This measurement permits electronic sequencing relative to camshaft rotation.

the current angular rotation of the camshaft and hence the location of the mechanically controlled devices on the machine.

Each electrically operated device on the machine can then be programmed to activate or deactivate at a specific degree of rotation of the mechanical camshaft. Accuracy will depend somewhat on program length and scan times, but accuracy of $+ 1°$ is typical.

Fixture Fault Memory. Information concerning assembly quality dimensions or functional characteristics can be carried not only in the processor memory but in the fixtures or pallets themselves through R.F. tag readers. The use of such tags is very valuable in free pallet systems, where pallets may leave the system, for repair loops or other reasons. Fixed index fixtures, permanently fastened in specific sequence to the indexing system, would usually rely on processor memory.

Somewhat akin is data recorded on the product itself by the use of bar codes. Date coding of production can be tied to specific batch lots of components if bar coding is utilized at the feeder locations to record batch numbers.

Indication and Documentation of Process Failure

Except for a random failure to insert or join, which is self-clearing at the next stroke of the machine, any failure of the assembly process on the machine will require corrective action by the machine operator. The sensors must be accurate. The operator should not have to look for the problem, but instead should have it pinpointed by some form of indicator light or digital language display. Frequency of failure together with resultant downtime is something that must be recorded and later analyzed. The display of these failure patterns can be as simple as a counter or can be sent to another computer running supervisory software. This sophistication is not required on the shop floor.

Input f,rom Sensors and Gauges. The qualities monitored in an assembly machine are for the most part displayed in digital form. Is the part fed properly or not? Is the switch opened or closed? Occasionally, however, the information is given in analog form, but the assembly machine needs the significance of this analog signal in digital form. Is it acceptable or defective?

The utilization of various inspection devices and probes is discussed in Chapter 8 in detail. It is sufficient in a discussion of assembly machine controls that the assembly system will ultimately require that the output of these tactile and nontactile sensors be supplied in a specific statement of acceptability or rejection, no matter how broad or narrow the range of acceptability or defection.

Indicator Lights. It is not enough that a machine stop upon detection of some problem. Operators must be quickly pointed to the area or nature of the problem if downtime is to be kept to a minimum.

Indicator panels are the most widely used system to pinpoint the problem. Digital displays expressing the problem in English text are widely utilized, but *whether this increase in sophistication will actually reduce downtime is very questionable.* Indicator light panels properly mounted are usually visible from any position around the machine. Experienced operators soon learn to equate a specific light or its panel position, even

at a distance or when viewed angularly, to a specific problem and a specific response. Whether digital language displays will have the necessary visibility to allow viewing from a distance or from an angle is most questionable.

With the exception of rigidly mounted sensors, such as electric eyes, it is quite normal for the machine sensors to advance to the workpiece during the dwell portion of the machine cycle to sense the presence, absence, or value of some condition. If the monitored quality or characteristic is absent, the output signal will either tell the machine to stop after this missing quality is discovered or after a predetermined number of failures have been detected. If the machine stops instantly upon detection of a malfunction, indicator lights will pinpoint the area of the problem, but the operator will be unable to take any corrective action since the machine will be stopped with all of the transfer stations and inspection units in their working position, thus securing the assemblies in their fixture nests.

To correct a problem, under these circumstances, it will usually be necessary to turn the INCH-RUN selector switch to the inch or jog mode and carefully jog the machine, if possible to the index portion of the cycle; that is, with all tooling and sensors withdrawn from the work fixtures so that whatever corrective action, be it repair, replacement, or withdrawal of the defective assembly, might occur. The selector switch will then have to be returned to the run position before the machine can be restarted. Precious seconds are lost at each failure.

A timing circuit can be incorporated into the machine control, so that the machine does not stop with the tooling advanced but will continue to run after the detection of a fault until the tooling is withdrawn from the work fixture. Machines with programmable controls can readily accept such holding circuits.

Arrangement of the indicator light panels should make them visually accessible from any position around the machine. They should be so arranged that illumination of any light will be readily noted whatever the ambient light conditions. Indicator light panels may represent the machine construction in plan view or be arranged in some convenient linear array. Panel arrangement simulating the machine's geometry may be beneficial to someone new to or unfamiliar with the machine, but operators routinely assigned to the machine usually find no problems with linear array and, in fact, may find it less confusing.

It should be noted that any such indicator light panel should segregate three types of conditions, either by linear arrangement or color of the indicator lights. The first condition would be those lights which indicate that the machine has been stopped by sensor detection of an assembly fault, which is essentially a product quality problem and does not indicate existing machine problems nor would machine damage occur if the machine were jogged to some other position for corrective action.

The second condition would indicate that the machine has been stopped because there is some indication of machine problems or problems with various services such as compressed air, inert gas supply, or some other condition under which any

attempt to operate or jog the machine could lead to severe damage or hazard. Usually illumination of such a light means that the machine cannot be jogged until corrective action is taken and proper condition restored by competent set-up or maintenance workers.

The third indicated condition is to reflect those circumstances where the operator has chosen to shut down the system because of some unforeseen or suspected problem. Strange noises, low transfer rail levels, or other problems not detected by sensors may lead the operator to stop the machine by the use of emergency stop buttons or lockout switches. The indicator panel should reflect this operator concern.

Indicator panels may show other conditions. Because of visual restrictions, sensors may determine low levels of parts in conveyors or transfer tracks that could indicate blockages or jams in feeders or magazines. Corrective action before the rail becomes completely empty may prevent

Fig. 4-11. Fault counters mounted on the panel box pinpoint insertion failures by specific station. They show incidence of failure, not downtime.

automatic shutdown of the machine. Some users feel that these indications should not be included in the main indicator panel but instead be mounted individually in a visually accessible position over the condition they are monitoring.

Other lights, usually larger, brighter, or blinking, or even audible signals may indicate that a seemingly inactive machine interlocked to other assembly or packaging operations by conveyors is in a triggered condition. This means that the machine, while apparently stopped, is in reality fully operational and will begin automatic operation of its own accord whenever conveyor sensors determine sufficient parts or conveyor capabilities are available for operation.

Counters. Most modern machine controls include some form of occurrence counters. Total machine cycles shown on a nonresettable counter may be the basis of preventive maintenance. Total cycles can also give some idea of machine

productivity. Total cycles complemented by the number of good parts made may indicate problems with product quality, machine behavior, or both. Most people, however, want to know the frequency of failure at each working station so that efforts can be made to identify major losses in net production before determining whether or not these problems are product or machine related. Data from counters such as those shown in Fig. 4-11 are a vital element of productivity audits discussed in Chapter 12. Most counters are resettable at the beginning of each shift, and their counts are recorded in appropriate logs before resetting.

Connection to Data Processors. Chapter 12, in its discussion of production audits, stresses that frequency of failure is significant relative to the extent of the downtime it causes. A station that fails 10 times an hour but can be restored in 15 seconds is not as significant at the end of an 8-

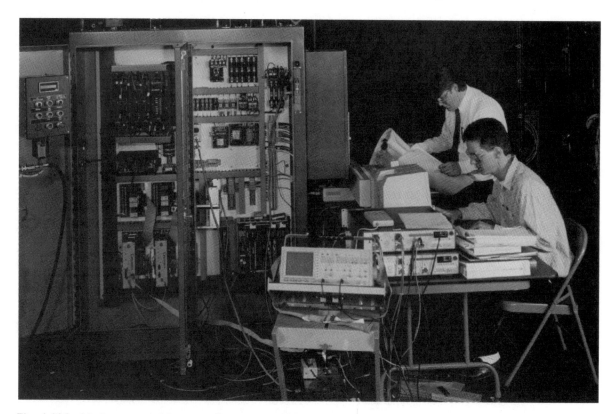

Fig. 4-12A. Modern assembly controllers are capable of providing a broad range of management information concerning production, quality and trend analysis.

Fig. 4-12B. Today's controllers provide shop floor management with significant production data, as well as expediting model changeovers.

hour shift as one failure that occurs in shift, but takes 40 minutes to correct. The first illustration would only take 20 minutes away from production or only half the downtime generated by the second example. Yet, if one were to go by fault counters alone, the first example would appear to be 80 times more serious: The recording of downtime as well as frequency of failure is essential to proper analytical resolution of problems.

The increasing use of advanced microprocessor based controllers is of great assistance in the collection of necessary information, since most modern processors have significant internal timing capacity as well as having counting capability. Extraction of that data can be through the use of permanent or portable programmers, or through data highway connections to central data processing computers. Many of the large microprocessor based controllers such as those shown in Figures 4-12 A and B are capable of data processing and can provide information in usable form to printers or computers.

SUMMARY

Throughout this book great emphasis is placed on the failure of the assembly process as being unfortunately a normal complement of assembly machine operation. Broad variations in product component quality that produce functionally acceptable products and machine-related failures will not be uncommon when one realizes that fixtures and transfer fingers, probes, and joining tools must accept significant part variations, locating small parts accurately and at the same time accepting the largest possible samples. Tooling for the most part will be essentially prototype tooling at time of machine delivery, and for most machines it will be one-of-a-kind.

Assembly machine control systems, therefore, must accept that occasional failures will occur, identify their location, facilitate correction of the problem, and record the necessary data so that problems in productivity can be assigned a priority and solved systematically.

Chapter 5
Feeding and Transfer of Parts

INTRODUCTION

This chapter is concerned with the methods and techniques of moving component parts of an assembly into the transfer and insertion stations on the assembly machine. Traditionally, unless fragility or cosmetic reasons interfered, most component pieces arrive at the assembly floor in bulk unoriented condition. The reasons underlying this include incompatibility between parts fabrication rates and assembly insertion rates, the purchase of components and fasteners from outside sources, the normal traditional organizational structure of a company in which certain fabrication operations are organized by specific technology departments, and lastly the need for secondary operations on many components after the initial fabrication operation.

Traditionally, therefore, those parts that can withstand the material handling problems without significant damage are transferred in bulk unoriented condition. This chapter discusses this traditional approach of presenting parts to assembly machinery.

It should be noted that increasingly companies are reorganizing plants into integrated manufacturing systems in which both the fabricating assembly and packaging machines are combined in a specific area dedicated to a product rather than a process. This change in plant configuration brings with it the potential to retain parts orientation established in fabrication and retain it entirely through the assembly process.

The advent of net shape production capability through improved tooling and imaginative product design has eliminated many secondary operations allowing a more direct coupling of fabrication and assembly. Additionally, integration of secondary operations or secondary forming work or parts fabrication directly on the assembly machine, which is discussed in Chapter 6, means that many parts can now come directly from the primary operation to the assembly system without going into bulk storage condition.

These factors in many cases dramatically reduce the number of parts feeders in many assembly systems. Increased emphasis is now placed on conveyorization systems with buffer storage capability and the use of palletized components in which fabricated parts are transferred directly into pallets that are used to feed mechanized assembly lines. An enormous growth in the use of pallets is now more technically feasible and financially feasible than at any time in the past. Such palletization may draw upon the flexibility of robotic systems for pallet loading and unloading. A discussion of such pallet systems is contained in Chapter 13.

Increasingly, assembly systems are being called upon to assemble components in which the matching tolerances are much smaller than can be controlled by normal machining processes. Fabricated parts must go through classifiers in order to sort a number of parts within the fabrication control limits into smaller increments required by the product function and assembly needs. These components, sorted into small increments, remain separated until the measurement of one component requires the feeding of a matching component of a specific dimensional tolerance. The increasing number of matched assemblies, particularly in the automotive component area, are further reducing the use of feeders designed to feed parts from unsorted bulk condition.

Despite these significant changes the majority of component parts that will come to assembly systems in the near future will still be of parts normally transferred to the assembly floor in unoriented bulk condition.

One basic cost reduction achieved by automatic assembly is through significant reduction in the manual handling of parts pieces and, in particular, a reduction in the cost of restoring part orientation to component parts, in bulk storage, that had some specific orientation at the time of fabrication.

This desire for lower assembly costs is reinforced by many important secondary considerations including reduced operator fatigue, elimination of physical injuries such as carpal tunnel syndrome, improved product quality, and safety. To obtain these objectives, automatic feeding of assembly components is one of the first aims of any assembly machine specifier. To translate this desire into concrete engineering, the assembly machine designer must make many judgments:

• Can fabricating operations be directly coupled to assembly systems? Are the component parts rugged enough to withstand damage from the tumbling actions of automatic feeders?
• Can insertion orientation of components be achieved through mechanical, electrical, or vi-

bratory means? Or must this orientation action be supplemented by human assistance?
• Is it more practical to have final insertion orientation performed within or outside the feeding mechanism itself?

The operational characteristics of the various types of feeding devices must be considered also. Vibratory feeders are usually quite gentle in their action and extremely efficient in their ability to orient complex parts. Elevator-type feeders can handle extremely large objects. Rotary feeders are both simple in construction and dependable in operation.

A prime design consideration is the reliability of the proposed feeding device: Can it meet required production or delivery rates? Does it have the ability to maintain these rates at various load levels? What is its resistance to wear?

Another important consideration is the extent of the availability of human labor at the assembly machine. While the ultimate goal is an unattended machine, self-correcting and continuous in operation, this ideal is not often reached. Human assistance is generally necessary at the assembly system to clear jams and to maintain hopper levels. If this help is available at the machine, it would seem wise to use this available help to facilitate difficult feeding operations. This is especially true when one marginal automatic feeding station may jeopardize overall machine efficiency and related production operations.

In theory, it should be least expensive to never release component parts from point of fabrication to point of assembly. In practice, however, there are a series of limiting considerations which must be weighed:

• Will the components require secondary operations, coating and plating operations, or heat-treating operations between fabrication and assembly?
• Will the pipeline costs of oriented part storage exceed the reorientation costs involved in feeding from storage?
• Will the fragility, lack of physical integrity, or cosmetic considerations of components preclude automatic handling and to what degree?

- Do the lubricants or parting agents used in fabrication mandate cleaning operations between fabrication and assembly?
- Will quality levels of incoming parts require presorting feeders, or other automatic or manual gauging operations prior to assembly?
- Do the fabricating rates, particularly where multiple machines, molds, or dies are involved, preclude direct coupling of fabrication machinery with assembly equipment?
- Will the maintenance downtime on fabricating equipment jeopardize assembly machine efficiency if the two machines are closely coupled particularly if done without buffer storage?
- Has the true cost of storage systems between fabrication and assembly been considered?

These are only some of the major points that must be considered. In a significant number of cases reliable part orientation and feeding is best done by manual operators, particularly if they are aided by ergonomically designed operator-assist units.

Available commercial technology generally supports the claim that almost anything can be fed automatically if someone is willing to pay for it. The price for automatically feeding may lie principally in the cost of developing the feeder, in the meeting of necessary feeder performance requirements, in the safe transport of parts by the feeder, or in some combination of these three factors.

Vibratory feeders remain a relatively new development in parts feeders. They were originally used to complement or supplement mechanical part feeders. Vibratory feeders, in many instances, can handle parts that could not otherwise be fed automatically.

Despite the competition of vibratory feeders, mechanical feeders are still a most practical solution to many problems faced in assembly machine design. At one end of the scale, elevator bin feeders handle parts too heavy for vibratory feeders. At the other extreme, simple fixed magazines are extremely low in cost and quite dependable. This chapter covers some of the more common types of mechanical vibratory and magazine feeders and outlines the important factors that influ-

ence selection of the most suitable methods of part feeding and orientation.

The role of robots in assembly widely heralded in the 1980's proved more controversial than articles in both technical and popular journals at that time indicated. For the purposes of this chapter, any robot can be considered an electro-mechanical transfer device. When viewed in this light, its functional versatility must be judged against other approaches in the light of its capital cost, difficulty of integration with the overall assembly system, cyclic capability, positional accuracy, and operational expense and durability.

In this respect, several points are worth noting. Julian Horne, Timex's widely respected Director of Automation, postulated in a major keynote address that almost all assembly motions require at most three axes of motion, well within the capability of most simple nonrobotic transfer devices.

Secondly, the feeding of component parts to an assembly from a bulk storage condition requires reestablishment of a positive attitudinal orientation before insertion into the assembly can be accomplished.

In conventional automatic assembly, establishment of part insertion orientation can be done in any one or more stages. It can often be positioned in the desired insertion attitude and radial orientation for final transfer by the part feeder. If this orientation is not possible in the parts feeder on a reliable basis without jeopardizing feed rates or reliability, there remain two additional areas to achieve the attitude required for insertion.

Assuming a feeder's ability to feed a component in any given attitude consistently and at desired rates (and without this possibility, feeding from bulk condition becomes most difficult), the machine designer may secure the desired final insertion attitude within the feeder discharge rails and/or related escapements, or in the transfer motion itself. A not uncommon situation may combine three stages of insertion orientation:

- Establishment of any one consistent attitude of part within the part feeder.
- Modification of that attitude within the feed rails and escapement.
- Final modification of the attitude to the desired insertional position in the transfer device.

Much of the robotic research conducted during the last decade was intended to duplicate human behavior in such an operation by having the robot utilize humanoid wrist and elbow movements to reorient the part from random attitudes through the use of visual and tactile sensor supplied intelligence. This "bin picking" approach has fallen to more cost effective alternatives.

PARTS FEEDING

Fabricated parts arrive at the assembly site in a multitude of ways. A most common condition is to have components arrive at the assembly areas in boxes or other containers in bulk unoriented condition. Storage density is usually excellent, and shipping costs minimal. Other parts, because of fragility, cosmetic considerations, warpage, or shipping costs are shipped in specific orientation-retention containers, such as trays, tubes, or mandrels. The means by which these components are extracted from these storage devices and placed in an attitude proper for insertion is the very heart of automatic assembly. It should come as no shock, therefore, to pinpoint that the most versatile of all part feeders remains the human assembly worker.

Hand Feeding and Operator-Assist Units

When management finally decides to mechanize assembly (and nothing happens without that momentous decision), it usually visualizes a sophisticated mechanism operating with little or no human intervention and minimal supervisory attention. In certain types of parts fabrication such as injection molding and press and screw machine operations, present state-of-the-art makes this condition practical.

In most discrete part assembly, however, the prototype nature of assembly system development compounded by continuous running product changes means that few assembly systems reach that point of machine maturity where they can be expected to run without assigned operators available to clear jams and restore the machine to automatic operation. Once one accepts the necessity of assigned labor for efficient, profitable operation of an assembly machine, and recognizes that a significant part of operator time

will be spent watching that machine run automatically, one should question whether or not that available labor can be better utilized to improve machine efficiency or reduce machine costs.

On any assembly machine there are always some stations that are less efficient than others because the characteristics of the part being fed or the insertion action do not lend themselves well to automatic feeding. Jams may occur frequently. In other cases, noise generated by part feeders may require such extensive sound enclosures as to have potentially serious negative impact on the system's net production. When either of these two situations occur, it can often be rewarding to provide for some form of direct labor integration on the assembly machine through ergonomically sound design. In some cases, operators may be able to feed these difficult parts directly to the work-holding fixture. In other cases some isolation of the operator from fixtures may be required. The use of operator-assist stations provides safety for the machine operators by isolating them from direct contact with the machine system. At the same time, the use of such assist stations provides a small degree of float (20–40 seconds) between the machine consumption and operator loading. This relieves the operator of the fatigue associated with matching each stroke of the machine and also allows them time to observe or correct other automatic stations while the machine remains in operation.

Figure 5-1 shows a simple rotary-type operator-assist station, while Figure 5-2 illustrates a short conveyor-type operator-assist unit combined with a fixed-receiver magazine. In many cases, operators can feed two different parts to an assembly machine through ingenious use of operator-assist devices and/or multiple-fixture nests.

Magazine Feeders

Magazine feed units are devices for feeding preoriented parts, one or more at a time, to a moving work fixture or subassembly. A simple magazine feed unit is shown in Figure 5-3. It holds a stack of oriented plastic pump dip tubes in a rigidly mounted receiver unit. Here the receiver chute is machined to the outside contours of the

Fig. 5-1. A small rotary indexing table contains orienting mandrels to simplify operator loading of vanes used on small bombs. After manual loading, the table indexes to an automatic transfer device for actual insertion of the vanes in the work fixture. Such an ergonomic device frees the operator to monitor automatic stations.

bars and the parts are hand oriented and loaded into place. Enough parts can be held in the receiver to run the machine for 10 or 12 minutes without additional loading. Hand-loaded fixed receivers of this type allow one operator to service many stations or several machines.

A fixed-receiver magazine requires frequent attention, and the parts must be loaded at the machine. To overcome this problem, replaceable magazines with loaded parts may be inserted into a fixed-receiver block (Figure 5-4). These replaceable magazines may be loaded in other areas and brought to the machine as required.

An important design consideration in selecting preloaded magazine feeding is the number of magazine tubes or inserts that will be required in transit between the fabrication/loading operation and the assembly machine. If the distance be-

tween these two operations is great, the number of magazines needed in transit can be extremely expensive and may prove to be detrimental to machine justification. The use of extruded aluminum or plastic magazines can help reduce this cost to a minimum.

In the fixed receiver type of magazine as shown in Figure 5-2, the machine can run constantly while fresh parts are manually loaded. In the replaceable insert type of magazine, it may be sometimes necessary to stop the machine to change the magazine inserts. Most machines are designed to stop upon detection of an empty magazine. To facilitate magazine changes on replaceable insert types of magazine feeders, turret holders commonly are used, as in Figure 5-5. Automatic exchange of replaceable magazines whether linear or rotary is quite expensive.

Fig. 5-2. This disc brake assembly machine optimizes operator utilization by an ingenious management of a fixed receiver magazine for friction pads and a short operator assist type indexing conveyor for the steel plates. The operator alternates between loading the magazine with preoriented parts and loading plates from bulk to the magazine.

Magazine changeover should occur without stopping the main system.

When the very nature of the part makes automatic feeding from bulk storage difficult, magazine feeding may be the only practical solution. Compression or spiral springs, for example, tend to tangle and interlock when stored in bulk. A multiple magazine feeder for very small fragile compression springs is shown in Figure 5-6. The magazine plates can be hand loaded or automatically loaded on a spring winder and then placed on the assembly machine. An indexing device moves the magazine plates, which have a spiral-hole pattern, toward the air pick-off tubes as the

plates rotate. In this way, the spiral-hole pattern passes under the pick-off tubes.

Magazine feeders can relieve the operator of the tedium of hand loading parts to the fixture, and often permits machine loading at a faster rate than would be possible manually. It also permits an operator to service several machines or several work stations.

Mechanical Feeders

Mechanical feeders that both orient and feed parts can be classified into two basic types. In one type, machined pockets or nests move through the bulk storage area and pick up a certain portion of the parts that are properly oriented. These oriented parts fall into the nests or pockets and are then carried to a discharge point where gravity causes them to slide down the feed track. The second type of mechanical feeder uses scoops, oscillating rails, or cleated chains to carry small portions of the stored parts over fixed orienting rails or chutes. The parts then fall by gravity onto the rails. Parts with the proper orientation are retained or captured while the others fall back into the storage area.

An early assembly machine, Figure 5-7, for valve tappets and adjusting screws uses both of the mechanical-feeding principles just described.

These two basic principles of feeding are applied in different ways in various types of mechanical feeders. Although nomenclature is not always consistent, mechanical feeders usually can be classified into one of the following types:

1. Elevator feeders
2. Blade feeders
3. Box feeders
4. Rotary selector ring feeders
5. Barrel feeders
6. Centrifugal feeders
7. Drum feeders

The first five of these seven feeder types are the most commonly used. Centrifugal feeders, however, have enormous potential for very high speed feeding applications.

Elevator Feeders. Elevator feeders, Figure 5-8, are used for bringing relatively large or heavy parts from bulk storage. The feeder has three ba-

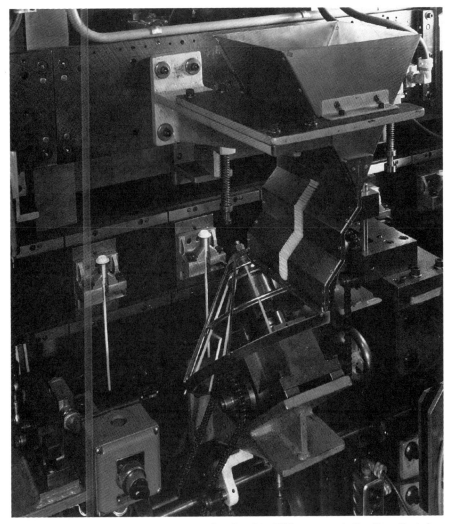

Fig. 5-3. A fixed receiver-type magazine feeder. This system is feeding dip tubes for plastic detergent pumps. The bottom part is wiped away by the index of the dial. A machine operator is responsible for maintaining the part level.

sic elements: a bulk storage bin, usually floor mounted; an elevating chain; and orienting and discharge rails. A floor mounted storage bin, which facilitates loading of large parts, has tapered sides that push the parts toward the elevating chain mechanism. A gate, which prevents parts from falling through the lower portion of the bin, has sufficient clearance to allow the elevating chain to pass through. The chain elevates the parts high enough so that orientation and selection may take place as parts fall downward over orienting rails or pass along vibrating rails. Improperly oriented parts are returned to the bulk storage area, while properly oriented parts are fed to the transfer unit.

The type of cleat fastened to the elevating chain will vary. In addition to the cleats used for elevating the parts themselves, rakes or other orienting devices are used to load the chain properly. As in any other type of feeder, the elevator feeder principle assumes that a certain percentage of parts in bulk storage can be successfully lifted by cleats on the chain and, of these, an acceptable percentage of elevated parts will fall on the orienting rails in a usable attitude and be captured.

Fig. 5-4. Replaceable magazine units may be preloaded elsewhere and rapidly changed in the lower fixed receiver when emptied. The fixed receiver provides a reservoir of parts during this changeover.

The noise of operation and the possibility of damage to parts by the movement of the cleated elevator chain, as well as the fall of the improperly oriented parts in the return chute, have limited the use of this type of feeder. However, properly designed elevator feeders can handle surprisingly fragile parts. Some feeders have been tooled to handle two different parts with the same elevating mechanism. For example, a circular cork gasket and a metallic lid used in assembly of oil filters are fed from dual bins by the same elevating mechanism and then are fed to adjacent stations on the assembly machine. The ability to

handle these cork gaskets illustrates the versatility of a properly tooled elevator feeder.

Blade Feeders. Modern blade feeders, Figure 5-9, use a bulk storage bin similar to those for elevator feeders. The selecting principle, however, is different. Here, tapered sides on the storage bin force the parts into the center of the bin where an oscillating blade, pivoted at one edge of the feeder, swings up and down through the storage area. As the blade rises, those parts that are readily oriented fall into a pocket in the edge of the blade. When the blade comes to the top of its

Fig. 5-5. Turret storage of magazines provides for automatic changeover of empty magazines. If exchange cannot occur within the index period of the machine, then a small fixed reservoir must be utilized to maintain operation during magazine exchange.

stroke, it is in line with the discharge track and gravity causes the parts to slide down through a gate into the discharge section.

In addition to the oscillating type blade, there are reciprocating blade feeders such as shown in Figure 5-7. In the reciprocating type, the top edge of the blade is shaped in the fashion of an inverted guillotine. When the blade is at the top of its stroke, the angled edge is in line with the discharge track. Blade feeders are quiet and efficient, but are limited in the main to flanged cylindrical parts or U-shaped parts.

Box Feeders. A box feeder is similar in operation to a blade feeder in that parts are picked up by a slot and discharged through a gate at the top point of the oscillation. However, in a box or, as it is often called, nail feeder, the entire storage area moves up and down, and the selecting slot

is machined into the bottom of the feeding device. Quite often, this type of feeder has several parallel selecting rails with multiple discharge tracks or tubes.

Both blade and box-type feeders are suitable for selecting only headed objects such as rivets or multiple-diameter cylindrical parts, which will hang from the head or largest diameter because of the location of the center of gravity.

Rotary Selector Ring Feeders. A typical rotary-type mechanical feeder, Figure 5-10, is tooled to feed small discs one at a time. Parts are dumped in bulk into the feeder bowl without specific orientation. A bottom plate, machined to fit the outside dimensions of the parts, circulates through the bulk storage area, wiping away those parts that fall into the milled nests. As this ring rotates, parts are carried upward until they reach a milled

Fig. 5-6. Cylindrical magazines provide dense storage and protection of individual parts not possible in vertical magazines.

section of the retaining gate. Here, one part at a time slips into the discharge rail. The milled pockets of the selector ring can be shaped to fit flat stampings, discs, cylindrical parts, and headed objects.

The selector-ring-type feeder can be made with a variety of configurations. Sometimes the nests are milled in the outer edge of the ring and parts are bypassed through the outer ring to a discharge point. Standard mechanical rivet feeders are usually of the selector ring type.

Another variation of selector ring units is the pin feeder, which is used to select cup-shaped objects, whose outside dimensions are tubular. Quite often, these parts are difficult to orient. The pins in a feeder are small probes that are cammed into the nests of a selector ring feeder. Properly oriented parts are retained by the pins, while improperly placed parts fall back into the storage area.

Barrel Feeders. Often similar in outside appearance to a selector ring feeder, barrel or drum feeders pick up parts from the bulk storage area with scoops or blades fastened to the side of a rotating drum. These parts are discharged by gravity onto orienting rails. Although the method of elevation is different, the selecting principle of this type of feeder is similar to that of elevator-type feeding devices.

Centrifugal Feeders. Centrifugal feeders are intended to sling parts to the outside of the feeder through the rotation of the center or bottom plate in the feeder. The motion imparted by the rotation of the center plate may be used to force parts into selector gates and rails for further transfer, by means of vibratory or belt-driven transfer motion. These feeders are very attractive for high volume applications from a theoretical standpoint, but have often proven disappointing in practice, since debugging is consuming of both time and sample parts, while the pool of technicians for such development is extremely limited.

When the center plate is rotated more slowly, these feeders are often used as reservoirs or accumulators in conveyor systems to relieve con-

Fig. 5-7. An assembly machine designed before the invention of vibratory feeders uses rotary and linear mechanical feeding systems.

veyors between two machine systems from surges that occur when one of the two systems is not operating. These accumulators serve as a modest, low-cost means of line balancing between two machine systems.

Drum Feeders. A drum feeder is distinguished from a barrel feeder only in the shape of its rotational bin. Many of the mechanical feeders previously described go back so far in time that their origins are unknown. For feeding certain parts, particularly washers, rivets, pins, and screws, they were efficient, quiet, and inexpensive. Some few configured parts had significant asymmetry coupled with proper location of center of gravity to permit their capture on the selector rails.

Until the invention of the Syntron Vibratory Feeder in the late 1940's, however, there was no good tool to automatically select and feed the broad spectrum of asymmetrical component parts used in high-volume assembly.

Before examining the vibratory feeder, however, we must mention a unique type of mechanical feeder—linear belt feeders—which had enormous but short-lived popularity in the mid-1980's. As vibratory feeders grew in size and complexity, so did the noise levels generated by their operation. Worker complaints, union grievances, and OSHA brought about modern sound enclosures. This is discussed in Chapter 9. In brief, however, the extensive use of sound enclosures greatly hinders assembly machine performance. Major customers sought relief from the dual problems of noise and damage. In the mid-1970's, a new approach to parts feeding was taken by several feeder manufacturers. Parts were spilled in a controlled way from storage bins onto conveyor belts and dragged by belt friction through a series of selection gates and orienting devices to an escapement. Belt life was a major hurdle in early designs. Additionally, the length of the earliest feeders impeded access to operating stations on

Fig. 5-8. A typical elevator feeder using a cleated chain to carry parts over a selector rail. Those parts with proper orientation are returned, while others drop back into the bin.

the machine. Figure 5-11 illustrates a modern belt-type mechanical feeder.

Many hybrid feeders are now evolving, combining the best characteristics of mechanical feeders with those of vibratory feeders. What percentage of the market these linear feeders, mechanical or hybrid, will take from conventional vibratory feeders remains modest. The large size and high costs of these feeders, as well as those of large centrifugal feeders, have greatly reduced their broad appeal. They do, however, provide advantages in carefully selected applications.

Vibratory Feeders

In the organization of this book, vibratory feeders come chronologically after mechanical feeders, but if mechanical feeders were the DC3's of early automation, vibratory feeders are the 727's of the automatic assembly industry. They are the

versatile, rugged, day-in, day-out workhorses that shoulder the majority of all automatic part feeding operations.

This section examines the use of controlled vibration for feeding and orienting parts. Controlled vibration can be used:

- To move nonoriented parts along a circular or linear path.
- To select properly oriented parts from those in an undesired attitude.
- To aid in actual orientation of improperly oriented parts.
- To transport oriented parts along a controlled path.

Figure 5-12 shows one of the simplest uses of controlled vibration—a vibratory operated bulk feeding chute. Powders, aggregates, symmetrical

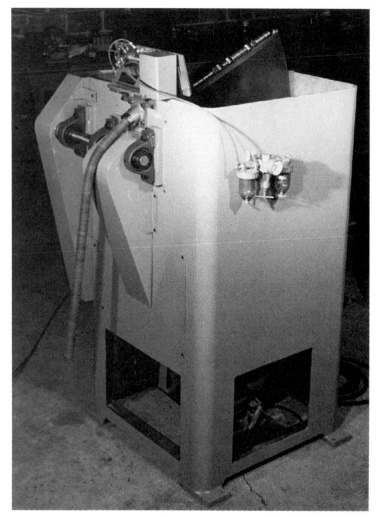

Fig. 5-9. A blade feeder uses a vertically oscillating blade to capture parts suitably positioned for discharge down a gravity rail.

objects, or nonoriented parts fed into the closed end of the chute from a storage bin are transported by vibratory action to the open end for loading, measuring, or mixing purposes. The vibratory action not only imparts a direction to the flow of the material, but also acts to loosen up and aid the flow of materials that tend to pack or interlock.

In assembly machines, it is usually not enough to move nonoriented parts. Generally, the requirement is to move individual parts from a bulk condition to an oriented condition for proper introduction into the assembly operation being used on the machine. Figure 5-13 shows a vibratory

feeder bowl in which a small-headed cup-shaped object is moved by the vibratory action from the center of the bowl upward and outward along a clockwise spiral track until it reaches the tooling section of the bowl. Here, only those parts that have the cup shape facing upward are allowed to move into the discharge track of the bowl.

How Vibratory Feeders Work. A typical vibratory part feeder, with the cover plates for the operating mechanism removed, is shown in Figure 5-14. This feeder unit is electrically operated. Today, most feeder equipment manufacturers use this type of vibratory electromagnetic motor sys-

Fig. 5-10. A rotary bowl feeder lifts parts and then drops them on discharge rails as the bowl rotates through the storage area of the feeder.

tem, although other methods have been developed for obtaining the desired vibration. In Figure 5-14, the bowl, which contains the parts, is mounted on a plate that can receive many other different types of bowls. These mounting areas are somewhat standardized, and allow many types of tooled bowls to be mounted on a common vibratory motor. In this instance, the bowl carrier plate is held suspended by three leaf-type springs that are fastened in turn to a very heavy base. The base is mounted on isolation pads so that most of the vibration generated in the feeder is confined to the feeder itself.

Also fastened to the base is an electromagnet. When the magnet is activated, the bowl carrier plate is pulled downward, causing the leaf springs to deform in the direction in which they are angled. When the magnet is deactivated, the springs snap back to their original position. As the springs return to their original position, the bowl is both rotated and lifted upward, causing the parts in the bowl to be thrust outward and upward. While the parts are still moving outward and upward, reactivation of the electromagnetic coil moves the bowl downward and, in relation

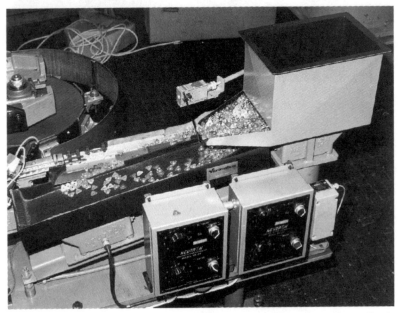

Fig. 5-11. Belt-type and brush-type mechanical and vibratory linear feeders greatly reduce noise and possible part abrasion by carrying component parts through selection gates.

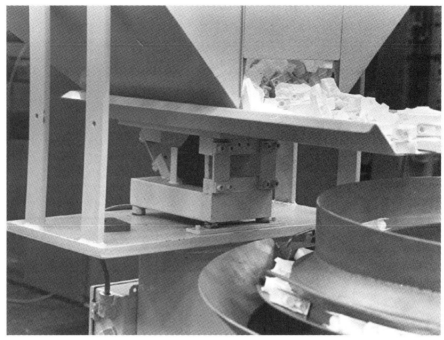

Fig. 5-12. A vibratory feeder used to feed bulk material along a specified path.

Fig. 5-13. A simple cup shaped part is fed to a selection section where scalloped edges reject parts coming with the cavity down and pass those parts with the cavity uppermost.

Fig. 5-14. Construction of a basic vibratory feeder. Larger feeders may use many more springs and several electromagnet "motors."

to the parts, backward. The parts then drop into a new and advanced position in the bowl. In this way, the pieces are advanced along a spiral track in the bowl.

The electromagnet is operated by alternating current passed through a selenium rectifier. Usually 110 V, 60 Hz AC is fed to the feeder control panel and is rectified into half-wave direct current. The amplitude of bowl displacement caused by the pull of the electromagnet is controlled by means of a rheostat across the power supply. When half-wave direct current is used, the impedance of the electromagnet coil tends to slow down both the build-up and drop-off of the current peak. In the development of vibratory feeders, considerable research effort in feeder control has been directed at obtaining a square wave current characteristic that will give better feeder performance.

The original Syntron patents have now expired, but some people involved in part feeding still use the term "Syntron" as a generic name for all vibratory feeders, whatever their construction.

In the United States, bowls using half-wave rectification vibrate at 60 Hz. In other countries, 50 Hz operation is common. There have been some who attempted to provide better tuning flexibility by modifying this 60 (or 50) Hz vibration rate. There were early but limited attempts at frequency-modulated controllers. These are rarely used. Others arranged control rectifiers in flip-flop fashion to produce a direct current turned on and off at 120 Hz (or 100 Hz) rather than 60 Hz (or 50 Hz) with some good results.

Warren Burgess of NASA developed a pneumatic drive system wherein the air motor adapted its vibrational frequency to the natural resonant frequency of the bowl together with its parts load. As the bowl emptied out, the natural frequency of the system changed and the motor compensated accordingly. Extremely good performance resulted, but pneumatic feeders have never gained broad acceptance in the marketplace.

Bowl Rotation. The feeder shown in Figure 5-14 produces a counterclockwise feeding motion. By placing the leaf springs in another direction the feeder can be made to move parts in a clockwise motion. In Figure 5-14, note that a mounting pad has been provided for the leaf spring to give bowl rotation in the opposite direction. Clockwise or counterclockwise movement of the bowl has no effect on the motion of the parts per se but it can affect the ability to capture or orient specific parts. This in turn can dramatically affect the machine layout and access to tooling. It is important, very early in machine layout, to determine with the feeder vendor if the originally selected rotation direction will harm feeder efficiency.

Resonant Vibration. For maximum performance, the feeder must vibrate resonantly. To achieve this condition, the feeder is "tuned." The tuning process basically consists of selecting the proper leaf springs or number of individual leaves in the springs and, in some cases, of affixing tuning weights to the bowl. Each tooled feeder bowl has its own vibration characteristics, and its resonant frequency will depend not only on the manufacturing variations in the feeder mechanism, but also on the size and type of bowl placed on the vibratory motor, the nature of the tooling fas-

Fig. 5-15. Auxiliary bins feed parts to the orienting bowl as required to maintain parts remaining in each bowl at a specific level.

tened to the bowl, and the present weight of the parts in the bowl. Thus, when a specific spring strength is selected for the feeder bowl, optimum vibration will occur at one specific weight of the material vibrated. However, as parts leave the bowl, this weight constantly changes, altering the vibration characteristics and performance of the feeder. Figure 5-15 shows an auxiliary bin feeder that is often used to help overcome this situation. This auxiliary feeder uses a vibratory discharge chute. A level-sensing float, which controls the auxiliary bin feeder, rides in the vibratory feeder bowl. As the parts leave the feeder, the float drops down. When a predetermined level of parts in the bowl is reached, the discharge chute of the aux-

iliary bin feeder is actuated. Parts from the auxiliary feeder drop into the vibratory feeder until the float rises enough to turn off the feeder chute. The high and low levels of parts loading in the vibratory feeder bowl are determined by the performance characteristics of the bowl, and are normally set at those points where a higher or lower level of parts would seriously affect the performance of the feeder.

In addition to ensuring high-performance rate of the vibratory feeder, auxiliary bin feeders have other important functions:

• To provide a much larger part storage than is available in the vibratory feeder bowl.

- To reduce to a minimum the operator attention required in maintaining a satisfactory part level in the vibratory feeder.
- To reduce feeder noise levels.

Servo Mechanisms for Feeders. The servo type of vibratory feeder control is another method used to keep feeder performance constant with a varying load in the bowl. In this type of feeder control system, a transducer, which measures the amplitude of vibration, is placed on the vibrating section of the feeder. As the amplitude of vibration varies with the changing weight of the vibrated bowl, this variation is electronically measured. The servo mechanism feeder control shown in Figure 5-16 adjusts the operating current to compensate for these amplitude variations, and, thus, the feeder performance is kept relatively constant across a wide variation in bowl loading.

Measuring Amplitude of Vibration. A simple indicator has been specially developed to measure the amplitude of vibration in feeder systems.

Figure 5-17 shows a typical installation on a vibratory-type feed rail. The indicator, which is in the form of a small plaque, is fastened to the vibrating bowl, and allows one to set optimum amplitude values that have been developed empirically. When the feeder or rail is vibrating, one of the three pairs of lines on the indicator will give the appearance of intersecting. Where this apparent intersection takes place, the amplitude of vibration can be read directly from the indicator scale.

Sometimes, special conditions may indicate the use of some input frequency other than the 60 Hz obtained with normal current. Special control panels and feeders that operate at other frequencies are available. For the most part, however, 60 Hz actuation serves the vast majority of all requirements.

Parts Orientation and Selection. The ease with which parts are oriented or properly oriented parts are selected depends on the configuration of the part and the location of its center of gravity. Some parts such as rivets or screws, which

Fig. 5-16. A servo controller controls the operation of a silicon-controlled rectifier to get optimal performance across a wide level of bowl loading.

Fig. 5-17. A vibration level indicator. Once popular, it is rarely seen today.

have a fairly large ratio of body length to head diameter, fall very easily into a slot in the vibratory feeder track and move along in a properly oriented position. On the other hand, a screw with a fairly large head and short body length may tend to assume a position with head crown and threaded body up because of the location of its center of gravity. When parts fall easily into one position, even though it is not the desired attitude, it is simple enough to turn them over to a desired position, either in the feeder itself or in the discharge track from the feeder. A more serious problem occurs when a part, either dimensionally or functionally, is not symmetrical but has a center of gravity that is located so that no predominant attitude is taken by the parts as they move up the vibratory track. For an overly simplified example, consider a rectangular part with six different faces that may face upward. If only one of these faces can be accepted in an upward position, those parts with any of the other five faces upward must be either returned to the center of the bowl or oriented properly. If experience indicates that each of these six faces takes the upward position in about equal frequency, and only one can be accepted, then the feeder actually must move six times as many parts as are required so that the number of correctly oriented parts can keep up with the machine cycle rate. In such cases, hopper efficiency becomes vitally important.

All of the various methods by which correctly oriented parts are selected and parts with improper orientation are turned to the proper attitude or returned to the bowl are beyond the scope of this book. Bowl development is a highly specialized art and the tooling of vibratory feeder bowls for this purpose demands much empirically derived skill, imagination, and experience.

In-house tooling of feeders on first installations is a needless risk. Advice on solutions to specific feeder problems can be obtained from many experienced feeder sources. Some feeder manufacturers have developed tooling manuals that show commonly accepted methods of orienting those part configurations most often encountered. Almost all manufacturers of vibratory feeder equipment will provide feeders tooled for a specific application. In addition, many firms specialize solely in the manufacture and tooling of vibratory feeder bowls. Finally, while many manufacturers of assembly machine systems are knowledgeable in the tooling of vibratory feeders, almost all system integrators leave the tooling of bowls to specialists in vibratory feeder construction.

Types of Vibratory Feeder Bowls. The earliest type of vibratory feeder bowl was cast aluminum. Bowls are classified by diameter, direction of rotation of feed (clockwise or counterclockwise), type of track (single, dual, or triple), and width and rise of the track. In addition to cast bowls, fabricated bowls, particularly of stainless steel, have gained the widest market acceptance. Fabricated bowls dominate today's marketplace. The advantages of light weight, less noise and ease of machining associated with cast aluminum must be balanced against the high wear rate often found in such bowls.

Because of the expense involved in tooling feeders, bowl wear can be extremely costly. Much of the wear on vibratory feeder bowls occurs when the discharge track is full and parts in the feeder continue to vibrate even though they cannot advance along the spiral path or the discharge chute. The parts then become small impact hammers and begin indenting the feeder bowl and track. Level-sensing units mounted on discharge tracks, which can turn feeders off when parts reach certain rail levels, materially reduce this type of wear, as well as help in noise reductions.

To reduce wear on feeder bowls, various types of coatings have been tried. Thin sheets of oil resistant neoprene, cemented to the spiral track, and various baked epoxy coatings are among the more successful methods used.

Fabricated stainless steel bowls are much more resistant to wear than the cast aluminum bowls. However, when fabricated bowls are used, care should be taken to ensure that tooling inserts are securely welded or fastened. Otherwise, vibration may loosen fasteners, or cause crystallization and fatigue of the bonding method. Today, most stainless steel bowls are heliarc welded and stress relieved.

Multiple Tooling of Bowls. When simple parts, such as nut blanks and threaded studs, are being fed, they can often be discharged into several feed tracks at the same time. Bowls with as many as seven or more parallel tracks can be used to feed separate discharge tracks from one common feeder. When multiple feeding is done, more than just the complexity of the tooling section of the bowl must be considered. Important factors include the type of transport rail necessary for moving the oriented parts, and the length of track required to span the distance between the hopper discharge point and the various machine stations where the oriented parts are inserted into the assembly.

In addition to feeders with multiple output tracks, installations have been made in which several selection bowls are mounted on a common vibratory base. Success of such installations depends greatly on the efficiency required of each feeder bowl, since there must be some compromise of the optimum performance characteristics for each of the bowls on the multiple bowl mounting.

Besides the multiple track and multiple bowl arrangements, other techniques have been developed for feeding several different parts from one hopper at the same time and even for assembling two or three different parts as they move through the vibratory feeder itself. The parts fed and as-

sembled in this manner usually do not require specific radial orientation.

Vibratory Feed Rails. When parts have been moved from bulk condition and consistently oriented, the next step is to move these parts to a transfer unit or station for insertion into the assembly. For simple parts, gravity feed along slanted rails may do the job. However. for most configured or thin parts, problems are encountered in attempting to use gravity feed rails. These problems involve not only the feeding of the parts but even the manufacture of the rails. Gravity type feed rails generally have an S-shaped configuration as shown in Figure 5-18. As parts cross the radiused section of the rail, clearance in the top retaining rail must be increased so that the parts may move freely. At the point where this extra clearance is provided, parts tend to climb on one another or "shingle," often jamming the discharge rail. The use of horizontal vibratory feed rails can reduce this "shingling." Such rails can also eliminate the problem associated with bending side and top rails so that they match uniformly, since there is no need for the S configuration. In Figure 5-19 vibratory rails are used to transport the output of a battery of three assembly machines to a common accumulator rail. Direction of flow along these rails is determined by the angular displacement of the vibratory springs.

Straight line vibration can be adapted for both feeding and orienting parts. Parts placed in bulk are vibrated through a selector gate that admits one part at a time to pass across a tooling area of the feeder in which properly oriented parts are vibrated to the discharge point. Here, improperly oriented parts are returned to the bulk section of the bowl. The shape of this feeder may aid machine layout. Manufacturers of such units claim that the difficulty in fabricating the linear tooling section is much less than that of matching tooling to a spiral section of a circular-type feeder.

Vibratory units are often used in conjunction with mechanical feeders in the orienting section of the feeder system. In Figure 5-20, a floor-type bin feeder with an elevator chain removes parts from the bulk storage area and elevates them

Fig. 5-18. S-shaped gravity rails were common in early vibratory feeder applications, but posed real problems in top rail clearance at the bend.

above a vibrating track. At the top of the elevator chain, the parts are dropped on to the vibrating rail and then oriented as they move along the rail.

Discharge Methods. Discharge methods other than gravity and vibratory type feed rails are used. Figure 5-21 shows two of these methods. The feeder at the upper right is coupled to the transfer unit by a close wound spring. The flexibility of the spring permits a direct coupling between the vibrating feeder and the transfer unit. When rigid transport rails are used, provisions must be made for an air gap between the rail and the feeder so that the vibration of the bowl is not damped by the rigid discharge unit. The second method shown uses commercial flexible nylon tubing, which is often very useful for containing oriented parts. The feeder shown in the lower half

of Figure 5-21 is representative of this technique. It uses a nylon tube discharge track coupled to a venturi system to transfer oriented parts to the machine.

Most importantly parts do not always have to move along the track in the actual attitude or orientation in which they are fed to the machine. Parts should be tracked from the feeder to the escapement in whatever attitude or orientation works best with the least jamming. The escapement device at the end of the feeder track and the transfer unit itself then can convert the best feeding orientation into a final usable insertion attitude or orientation. The role of escapements will be discussed in the next section of this chapter.

Vibratory feeders seem to have practically unlimited application. Their ability to handle most fragile parts without damage provides a wide range of potential uses where the jostling effect

Fig. 5-19. Horizontal vibratory rails are used to transfer oriented parts to a transfer unit, sometimes changing parts orientation.

of mechanical feeders cannot be tolerated. The art of tooling up these feeders has become highly specialized, and some new type of feeding problem is being resolved almost daily, by specialized feeder vendors, building on a broad diversified experience base.

Selecting Vibratory Feeders. While it is most strongly recommended that the tooling of vibratory bowls be left to specialists, it does not at all mean that selection of vibratory feeders and their construction details be left totally to bowl builders. The machine designer and the machine user both have vested interests in selecting the feeder type, size, and inclusion of features that simplify operation, and aid in obtaining the highest possible net production by providing machine attendants with clear easy visual and manual access to the feeders.

As mentioned earlier in this section, there are several actions that occur within the vibratory feeder: movement from bulk into a specific path or track, preliminary selection and passage of those parts with desirable orientation, moving the preliminary orientation of components into a usable orientation, and transportation of these oriented or selected parts along a controlled path to the transfer point.

In the earliest days of vibratory feeding, parts were fed in cast aluminum bowls. Urethane liners, spring steel liners, and epoxy coatings were used to reduce bowl wear and part contamination from contact with aluminum oxide. These bowls were relatively quiet.

Since cast bowls were small, the internal tracks or ramps had to serve the requirements of feeding, selecting, and orienting in a small area. Even with multiple turns of the spiral ramp, selection

Fig. 5-20. A vibratory rail selects properly oriented parts after rough feeding is accomplished by the elevator.

and modest orientation had to occur within a relatively few inches of part travel. It is not unfair to say that early bowls were limited to parts with complete symmetry or significant asymmetry. For the most part, these bowls, referred to as "cascade" bowls, with internal selection and capture devices, did little real component parts orientation, but attempted to select or capture only those parts coming up the ramp in a usable attitude for transfer and insertion.

It was soon apparent, when vibratory feeders went far beyond the types of parts normally fed in mechanical feeders and were applied to parts having relatively slight asymmetry or minor imbalance or more complex outer geometry, that selecting parts in the cast upward spiral ramp with a usable attitude was not always very practical. It was often necessary to abandon the first attempt to select a desired attitude, and instead choose any one attitude that could be selected readily inside the bowl. Modification of the selected attitude to the desired attitude was then accomplished in the transfer tracks, parts escapement (of which more later), or transfer unit.

In truth, early part feeders for the most part selected individual pieces having a usable attitude and rejected or spilled all others back into the bowl. The emphasis was on selection with a very limited amount of orientation possible in the limited track length available and under the duress of back pressure from other parts climbing the rail.

As the push increased for better feed rates and for greater ability to select more complex shapes, early emphasis was placed on improving the driver; 120 Hz vibration, air motors, independent spring suspensions for vertical and horizontal motion, the use of silicon controlled rectifiers, greatly aided but did not solve every problem.

There was a parallel and ultimately far more important movement to improve the technology of bowl tooling. The great breakthrough occurred when the concept of "outside-the-bowl" tooling was developed (Figure 5-22). Internal bowl tracks were used to convey parts upward to the top of the bowl and then to external selection and orientation tracks on the outside of the bowl, which were either horizontal or fashioned in a downward path. There was no real practical constraint on the length of the selection and orientation section of these internal tracks, and parts then could be maneuvered into the desired attitude for transfer both within and from the feeder. This outside tooling capability has been fully exploited, often reducing the need to utilize complex escapements and transfer devices to achieve final insertion orientation.

The real design question for the assembly system designer and ultimate customer is the wisdom of depending totally on the feeder for final orientation of the part. A serious byproduct of this outside tooling concept has been major increases in feeder size resulting in greater operational noise levels and a significant difficulty in operator access. Noise now comes not only from the sound of parts meeting with one another, but in noise generated by the very movement of large feeder bowls moving in the air. The normal chosen solution to this noise problem is usually the use of sound guards, reducing both visual and manual operator access so often necessary to maintain system productivity.

This necessity to have complete operator access is because the occurrence of jams may become

Fig. 5-21. Flexible tubing or closely wound springs can be used to transfer parts from feeders to escapements or transfer units, while other parts can be transferred and reoriented end-for-end in flexible tubing.

more frequent as the complexity of orienting, selection, and confinement of component parts increases when demands are placed on the feeder bowl to present the part in the final insertion attitudes, rather than use escapements and transfer devices for such final positioning.

There is a very fundamental design question about the practicality of extremely large feeder bowls. The very size of such bowls, even without sound guards, makes parts loading difficult and jam clearance frustrating for operators of short or normal stature.

Some purchasers institute rigid procurement specifications for vibratory feeders mandating prolonged periods of parts storage within the bowl. Inevitably, this specification leads to increased downtime.

As a general design rule, bowl size should be restricted to the smallest possible diameter consistent with the tasks of the feeder and the machine. To fully determine this matter, selected bowl fabricators should be queried whether some other part attitude than final insertion orientation could be reliably attained in a smaller bowl.

Reduction in bowl size, possible elimination of noise guards, significant reduction in jamming, and ease of bowl loading are potentially valuable cost reductions and improvements in operating efficiency that may be available through greater dependence on escapements and transfer devices for final insertion orientation.

The second area where system designers and ultimate machine users may wish to invade bowl development is to ensure that clearing jams is

Fig. 5-22. Outside-the-bowl tooling has greatly increased the efficiency and capability of vibratory feeding techniques. Upward tracks inside the bowl elevate parts to the selection and confinement tracks outside the bowl.

made as easy as possible. Avoiding jams in the first place is an obligation of the bowl builder, but the bowl fabricator can only do this job well if he or she has a full exposure to the actual production component parts as they will come to the assembly machine. This is usually not always easy or possible, particularly in a world of simultaneous engineering. It has been pointed out throughout this book that today most assembly machines are purchased to assemble a new or significantly modified product.

Real production samples may not be available at the time of machine development. Feeders usually have to be developed using prototype parts. It is to be expected, in this situation, that feeders may require further development after production commences using real production parts on the user's floor.

In order to reduce, so far as possible, tendencies to jam in the selection, orienting, and confinement areas of the feeder tracks, the bowl builder must have exposure not only to the full variation in usable components, but to the defective parts and foreign material that will be routinely found in the containers used to deliver parts to the machine. The designer (and customer) may assume erroneously, and to their regret, that the bowl builder realizes that such contamination is routine in the user environment. In a recent discussion of startup problems on a new machine, a customer's engineer complained the bowl builder should have known that broken parts and foreign material would often be mixed in the supplied samples. They felt the bowl builder should have made provision to separate these defects automatically as part of the project. Most bowl makers would not agree.

No matter how high the management level at the time of system proposal and concept, execution remains at modest operational levels. The system development work is usually done by the people low on the corporate totem pole. It has been pointed out elsewhere that tool designers so often concentrate on how to make the machine work that they may overlook or neglect designing into the machines the necessary facility of correcting problems when it does not work.

Bowl builders are artisans, working in a hands-on empirical environment; watching, analyzing, trying out various approaches to get the feeder to perform its primary function, that of feeding acceptable parts in a specific orientation.

The machine builder and the ultimate end system user must be certain before accepting a vibratory feeder that the bowl builder has tried, within the limitations of the available sample parts, to ensure that the bowl will handle all of the production variations found in usable components and will reject either to discharge chutes or to the inner bowl without jamming broken or deformed parts and normal foreign material, such as sprues and rod ends.

If the customer and builder find that they are too busy to make this capability determination before the bowls are accepted, they will be forced to find plenty of time in debugging and installation to focus on this problem.

JAMMING. Jams in a feeder bowl will occur in three principal areas: the initial selection areas, the initial confinement area, and the transfer tracks. Jams in the selection area and in the transfer point of feeder to track are cleared in short order. A few seconds and human fingers are usually sufficient once the problem is detected. Before the advent of sound enclosures, a conscientious (and educated) operator would spot these jams by observing part levels in the transfer tracks. Without this visual access, photoelectric sensors may be required to notify, through some signal, unusually low track levels resulting from jams. It is imperative to remove these jams before the transfer track completely empties to avoid delays in restarting the machine.

A much more serious type of jam is that which occurs when a component part (or foreign material) jams along the transfer tracks. Removal of a side or top rail to remove the jammed part will mean excessive downtime if no provision has been made to facilitate the removal of jammed parts in the transfer track and to restore track rails to proper position. There is no more fruit-ful area of ensuring acceptable levels of performance on the installed machine than to find ways at time of feeder development in which to facilitate removal of parts wedged in the transfer rail. If at all possible, eliminate the top rail; utilize quick release devices, gates, or escape slots. The wide variation in rail design and product shape will not allow specific rules, but finding solutions is imperative. Again, it is an area best discussed with the feeder builders before they begin their work. *Do not assume* that bowl builders feel defective parts are their problem. They may feel with much justification that it is a customer problem. In this case, an ounce of prevention is worth far more than a pound of cure.

SORTING FEEDERS. The above topic leads to the question of sorting feeders. Quite often, and particularly since the advent of outside bowl tooling, continued development of a feeder bowl may provide some degree of parts sorting ability as part of the normal function of the feeder. Cooperation and patience (and paying a fair price) will be most useful in determining how much of this can be accomplished during bowl development and before final bowl welding.

In some few cases where high productivity is required on the assembly machine, and downtime prohibitive, it may prove necessary to pre-sort parts before placing the parts in the final assembly machine feeder bowls. Some few customers even request that sorting bowls be furnished with the machine. *Long industry experience would indicate that procurement of sorting bowls prior to gaining operational experience is neither wise nor practical.*

In order to develop any parts sorting bowl, it is necessary to know the types of defects and foreign material to be sorted and the extent or prevalence of each type of problem. If the problem can be solved by automatic means in the sorting bowl, it could probably be done in the primary feeding bowl, particularly if the problem is not an excessive one. It should not be expected that sorting bowls will never jam or that they can handle every type of sorting problem without human intervention.

Spring and Part Separators. No discussion of part feeders is complete without some discussion of parts that tend to adhere to one another or tend to tangle severely. In some cases, these parts defy feeding from bulk and may have to be fabricated on the assembly machine or be magazine fed or hand fed to the machine or conveyor.

Springs are notorious for the problems they present to vibratory feeders. Several developments in recent years have helped in many cases. The first is the use of centrifugal separators. A conventional vibratory spring feeder is designed to pass through to transfer those springs that are not tangled, but to divert tangled springs into the top of a constantly rotating parts separator. Such a separator, illustrated in Figure 5-23, has a series of paddles or lobes that strike the tangled springs. As the springs are compressed and then return to their free state, they often become untangled. They then are returned to the main feeder. Some springs never untangle. Every so often it is necessary that the separator be manually cleared.

A centrifugal separator of this type can tend to damage, distort or even break some springs. Each potential application will have to be tried empirically to determine if the spring separator will solve its problems.

Fig. 5-23. A cylindrical housing encloses an oscillating agitator which strikes tangled springs. The compression of the spring caused by striking the agitator often untangles the spring.

A second approach to spring feeding is the use of a European development in which small springs are kept in constant motion by currents of air. This constant tumbling motion seems to be very effective in separating springs. Loose individual springs passing a suction tube are pulled out of the storage area to the transfer point.

Parts that adhere together may do so because of accumulated static charges. Some experimenters report success with blowing a stream of ionized air into the bowl. Others question the environmental wisdom of having an operator work in this environment. Increasing prevalent environmental concerns make this approach a matter of concern.

Rubber parts are very susceptible to heat, cold, and humidity. Rubber parts that become so deformed in storage may often return to original molded condition by placing a chromolox type heating element over the bowl and thus bring rubber temperature into the range of 150°F (65°C). This heating also relieves stickiness. Rubber parts also tend to feed better when dusted lightly with talcum powder.

Ferrous parts that cling together may be separated at the escapement by subjecting them to a strong magnetic field. Parts acquiring the same charge repel one another and the uppermost parts pieces can often be captured by vacuum pick-up heads.

Metering and Dispensing

Automatic metering and dispensing operations are commonly associated with canning, bottling, or packaging machinery. However, many small mechanical and electrical assemblies require some powder, fluid, lubricant, or adhesive, which must be metered, dispensed, or applied to the parts. For such operations, assembly machines differ from most canning and bottling machinery in that the parts being assembled generally must have a specific orientation and position in relation to other parts so that the assembly may function properly. Also, assembly machines differ from packaging machines in that the transferred components generally perform some function after being assembled, and may require some form of functional inspection.

In this chapter, attention will be focused on those metering and dispensing methods suited primarily for incorporation in assembly machines. Those techniques associated mainly with canning, bottling, or packaging operations are not covered here.

Closely allied to the problems of metering or dispensing fluids, solvents, adhesives, or liquids for assembly operations is the question of the assembly machine's own lubrication system. The same techniques used for one often apply to the other. Hence, the same control and design considerations are equally applicable in each case.

Automatic Grease Dispensing. Many small assemblies require the application of grease during the assembly process. Some of these assemblies will not have further lubrication during their service life. Therefore, it is important to the service life of the assembly that the grease be applied at the right point and in the right amount initially. Typical examples of such requirements are electrical wall switches in which grease is applied to retard electrical arcing as well as to facilitate the mechanical operation of the switch itself. Sealed ball bearings, automotive brake adjusting screws, window raising mechanisms, and other items that normally would not be lubricated after assembly also illustrate types of products that need some lubrication during the assembly process itself.

When a switch from manual operations to automatic assembly procedures involving automatic lubrication is anticipated, several considerations should be kept in mind. Usually, the same type of grease pumping equipment that is used in automotive service stations functions ideally for supplying lubricant greases, under pressure, to assembly equipment. Since there may be some distance between the pumping unit and the metering or dispensing equipment on the assembly machine, it may be necessary to determine the flow characteristics of the grease to be dispensed. This feasibility is determined by viscosity and flow rates.

Problems that arise in feeding lubricants automatically during assembly operations generally are the following:

- Controlling the amount of lubricants.
- Preventing contamination of the assembly machine and its fixtures.
- Ensuring that the lubricant is placed in the proper area.

Proper design of the metering and dispensing unit can usually solve these problems. The simplest approach is to have the metering unit dispense the lubricant only upon physical contact with a properly positioned workpiece; see Figure 5-24. Here, the grease pumping unit keeps grease under pressure for feeding to the metering device. When the vertically traveling tooling platform moves downward, the metering device is tripped by physical contact with the workpiece. If an empty nest in the workholding fixture comes underneath the metering unit, no grease is released.

Another method for preventing discharge of grease into an empty workholding fixture is to incorporate an inspection probe for checking proper presence and position of the workpiece in the station immediately preceding the lubrication station. Grease discharge is then controlled by solenoid actuation of the dispensing unit, according to the signal from the probe.

Fig. 5-24. This greasing station will dispense only if a part is present in the work nest.

Release of grease or lubricant into an empty workholding station will not only cause contamination of the machine and future assemblies, but, in lighter and more fragile assemblies, can lead to a serious reduction in the efficiency of the various transfer units because of adhesion of the workpieces to the transfer fingers. In addition, release of lubricants when no assembly is present to receive them is a needless waste of material.

Automatic Machinery Lubrication. High-speed assembly machinery needs proper lubrication to function efficiently. Too much lubrication can be just as serious a threat to efficient assembly machine operation as insufficient lubrication. Machine lubrication systems (Figures 5-25A and B), whether designed for oil or grease, have four basic elements:

1. A positive-displacement pumping arrangement for delivering the lubricant under pressure from a reservoir.
2. A distribution network for bringing the lubricant to the bearing points on the machine.
3. A system for timing the dispersal of the lubricant.
4. A metering device that controls the amount of lubrication applied to each of the bearing points.

Within this general framework, there are varying degrees of complexity, depending, generally, on the extent of safety features incorporated in the

Fig. 5-25B. A central lubrication controller.

Fig. 5-25A. A lubrication metering block contains flow to various lines.

system used. The chosen system may consist of nothing more than a distribution system spreading out like the branches of a tree from a common trunk. Or, the various metering points may be interlocked so that the distribution of lubricant to each bearing point is dependent on the previous metering of some other point in the system. Finally, the system may be interlocked so that blockage of any distribution point or distribution line will create excessive pressure within the system, actuating a safety valve. This valve, in turn, may trigger visual or audible warning signals of lubrication system failure or may, by electrical interlocks, shut down the assembly machine until the malfunction is corrected.

The danger of overlubrication is greater on assembly equipment than on fabricating equipment because of the possibility of contaminating the workpiece, delicate transfer mechanisms or even clean room environments. In practice, many of the lighter mechanisms on centrally lubricated or lubrication free machines are manually lubricated for reasons of cleanliness. When the design of the machine dictates this method of lubrication, special attention must be given to the selection of the materials used in the transfer devices. Hardened steel with cast iron and sintered bronze on hardened and ground tool steel are excellent combinations that require minimum lubrication. Another excellent combination is hardened tool steel and anodized aluminum. Synthetic lubricants (without a petroleum base) are available for selective application. New surface coatings for sliding and rotating parts eliminate much of the need for lubrication. Increasing use of clean room environments for manufacturing is directly responsible for new developments in this area.

Feeding Adhesives and Solvents. The principles of high-speed application of adhesives are much the same as those used in inking printing presses. The adhesive is first placed on a transfer pad or roll for subsequent application to the workpiece. This type of application is usually more suited to continuous motion machines than to intermittent motion devices.

The problems of precise metering and contamination of the assembly machine are compounded by the special requirements for feeding solvents or adhesives. Usually these materials will require a certain time period to permit setting or bonding to take place. Since assembly machines run at a fairly high rate, the short time interval between the joining operation and the ejection of the completed assembly must be considered. The nature of ejection may have to be altered because of the short curing interval and the need for maintaining a proper attitude or position of the parts during the curing operation. Also, a necessary bonding or curing time interval may limit the productive capacity of the machine or may even dic-

tate the use of a longer or larger machine to allow enough time for initial adhesion or joining to occur. System designers need to distinguish between full bonding strength and the time when sufficient bond exists for free ejection. Means are available to accelerate curing times by the use of chemical accelerants and ultraviolet light.

These problems are not always easily resolved. However, their existence cannot be overlooked when determining the potential of automatic assembly methods for this type of joining.

In the machine shown in Figure 5-26, it was found that when accidental discharge of solvent from the metering pad type of applicator occurred, the machine had to be shut down to permit cleansing of the dial and work-holding fixtures. A modification in the design of the component, in which a small flange was added to retain the solvent during the joining operation, eliminated this condition and greatly increased

Fig. 5-26. The use of a metering pad approach to solvents, ink or oil, is inexpensive, but may tend to cause contamination after prolonged machine stoppages.

the efficiency of the machine. Quite often, as in this case, the design of the machine and the efficiency attained is greatly dependent on specific product design modifications to permit efficient automatic assembly.

One excellent source for determining metering units for epoxies, adhesives, and other bonding or sealing agents is to seek technical assistance from adhesives manufacturers. They can often provide metering units with flow adjustment controls, purging systems, and disposable storage reservoirs. One should not be reluctant to use disposable containers and syringes in such a system. These are acceptable "perishable tools," which eliminate many operational problems.

The wide scale use of calculators, printers, imprinters, and like items has brought into play a variety of disposable products requiring the metering of a known amount of ink at high speeds. The use of peristaltic pumps is a wise one on assembly machines for these products. A peristaltic pump connects a fluid through flexible tubing to a discharge orifice or nozzle. This tubing is then compressed by a series of rollers mounted on a wheel driven by a stepping motor. As these rollers depress the tubing, they push fluid in the tubing forward to the orifice. With a stepping motor it is quite easy to adjust the number of tube compressions generated by the rollers as the wheel rotates a specific number of degrees.

It should also be noted that the flexible tubing, while perishable, is quite inexpensive. This type of pump makes changeover from one color to another extremely simple, since the ink reservoir, tubing, and discharge nozzle can be entirely replaced in a minute or two without the need to clean out or wash the pump mechanism.

THE ROLE OF ESCAPEMENTS

It was mentioned earlier in this chapter that when a part piece is being fed from a bulk unoriented condition to a final attitude or orientation from which it can be inserted into an assembly, there are three opportunities to obtain the final desired insertion attitude: part feeders, escapements, and transfer devices. It was also mentioned that many complaints about vibratory feeder performance find their basis on an erro-

neous machine design decision, namely, to have the feeder do all of the necessary orientation within the part feeder and its discharge tracks. Parts fed in any feeder will tend to orient in certain ways which may or may not be optimal for the feed ramps and selection tracks. Feeder vendors may flatly state their total inability to feed parts in a specific insertion orientation. When they take this stand, the machine designer has little recourse but to accept some other attitude, or orientation, which the feeder source can reliably achieve, and then convert the feeder orientation into a usable insertion attitude by means of tooling design ingenuity in the escapement or transfer unit.

The problem is clear cut when the feeder vendor has full confidence in his or her ability to feed a part in the desired final insertion attitude. The problem is additionally straightforward when feeder sources state bluntly that they cannot supply a feeder capable of feeding a part to a desired orientation. The real problem occurs at the concept design stage on those parts that seem marginally capable of being fed in a directly usable state.

The machine designer should, in this instance, act on the basis of the worst possible situation. The system designer and the customer project manager must realize that the relatively low price for part feeders and the competitive nature of the vibratory feeder business put great pressure on most feeder sources to complete the feeder and ship it at the first available moment.

The machine designer and customer must also recognize that in determining whether or not a feeder is acceptable for shipment to the systems integrator, the feeder vendor will run performance tests in which the parts flow from the feeder in a continuous and free-flowing fashion. In actual production practice, however, feeders do not operate in that mode on an assembly system. Parts coming from a feeder usually are fed into a transfer track or tube. Unless the assembly machine is designed (and this is quite rare) to wipe the parts away during fixture index, the parts coming to the machine in a transfer track must come to some stopping point (escapement) whence they are picked up or stripped away to be placed in the workholding fixture or assembly. Once the outermost part is stopped, subsequent parts coming from the feeder back up with increasing pressure applied to computer parts already in the transfer track. These parts pushing against one another move intermittently forward the length of one part piece as the outermost part is removed for transfer and insertion. In the meantime, the accumulated parts continue to push and jostle against one another. It is the feeder performance when parts are in this condition that is the significant one for system designers and users. When it is difficult to obtain a desired insertion attitude in a feeder, but it is decided to build it this way anyhow, one may find a feeder to be reasonably efficient in the free-flow mode, but marginally unacceptable in the intermittent flow normal to an operating assembly system. Under such marginal conditions the designer should accept whatever components attitude feeds most reliably and use escapements and transfer units to attain final insertion attitudes.

The stopping of components at the point of transfer unit pick-up for insertion is a major design challenge in the development of an assembly machine. This is true whether the parts are coming to the assembly system from a vibratory feeder, from a mechanical feeder, or from a buffer storage or transfer conveyor. The first part in the escapement mechanism must be properly positioned for pick up. The escapement tooling must be sufficiently strong to withstand the cumulative pressure of all of the parts pushing against one another in the transfer track or conveyor. It may be necessary to isolate the part to be picked up from contact with following parts. It may additionally be necessary to modify a part's attitude or orientation from that presented by the feeder track. These vital functions are done by escapement devices.

There are several levels of complexity of part escapements, depending on the functional role the escapement is to play in the machine system.

If the escapement devices are required merely to stop the forward motion of parts without any obligation to isolate the part or change its attitude or radial orientations, a simple "dead end" may be sufficient. The "dead end" illustrated in Figure 5-27 is simply a rigid block at the end of the active feeder rail; a block machined to accept

Fig. 5-27. A "dead end" escapement. Such an escapement is passive. It accepts a part for pick up without isolating it from end pressure or any ability to change the attitude of the part.

one part piece accurately for pick up, transfer, and insertion. This type of escapement must accept the largest possible acceptable part, while at the same time providing sufficient locational accuracy for the smallest part fed to it. The usefulness of a "dead end" type of part escapements is totally dependent on the freedom that the part ready for transfer has to be lifted from the dead end block without being damaged by or causing damage to the next adjacent component part, which is pushing against the part being transferred.

In the next order of escapement complexity, one may isolate the transferred part from the pressure of subsequent parts by the means of spring-loaded detents. These detents are movable stop jaws with sufficient spring preloading to hold back all of the parts in the transfer rail. The transfer unit then strips the part to be transferred past the detent jaws, rather than lift the part out by an upward motion. The jaws then snap back to retain the next part and all subsequent parts.

This is the simplest form of back-pressure isolation.

If it proves that spring-loaded jaws are not strong enough to hold back the pressure developed in the transfer rail, a more positive form of isolation is the shuttle escapement (Figure 5-28). Here the retaining jaws are mechanically, electrically, or pneumatically actuated. In another variation, a receiver block, instead of being rigidly mounted, is fastened to a slide. The slide then shuttles to the side freeing the part in the receiver from the pressure of parts following.

If the part to be fed must not only be isolated but have its original attitude modified, then the escapement may perform one of two additional functions. A change in attitude may mean that the planes of the part or its surfaces must be angularly rotated. For example, the surface on the top may be rotated until it is on the bottom. This can be accomplished by pivoting the shuttle to the desired angular position rather than by sliding the escapement shuttle horizontally.

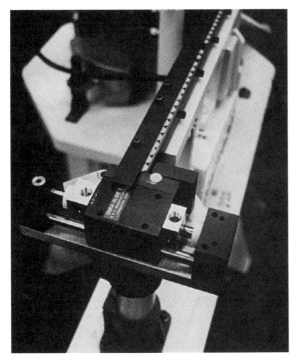

Fig. 5-28. A shuttle-type escapement is used to isolate parts to be transferred from subsequent parts in the rail.

A second type of positional change may concern modification of the radial orientation of the part. Here two different concepts may be required. It may be that, because of some locational feature on the part, it can be fed with the proper surfaces uppermost, with a consistent but uninsertable radial orientation. Here a simple escapement would suffice (dead end or shuttle) with final radial orientation of the component accomplished by a pantograph motion in the transfer unit itself. This approach may often be required not because the feeder is incapable of obtaining proper component orientation, but because the feeder and its associated transfer rail cannot be placed, owing to machine layout problems, in the proper position relative to the machine and its fixtures. This does not involve a major shuttle design, but rather uses the axial versatility of most transfer units.

A different type of radial orientation activity is found when the features requiring radial orientation, such as a hole or slot in a cylindrical part, are not external; do not cause major external asymmetry; or do not have a significant bearing on the location of the center of gravity. For cylindrical parts lacking one or more of these attributes, final orientation within the feeder is virtually impossible and, even if attained, is difficult or impossible to retain in the transfer rail. Here the escapement must accomplish two functions. The leading part must be shuttle escaped to a radial orientation device that will engage the important feature in the part and rotate it to a desired angular condition. If engagement is impossible, the part may be rotated until mechanical stops engage the desired feature. Here the escapement performs a role not practical in the feeder or transfer unit. Figure 5-29 illustrates a shuttle incorporating positional modification.

PART TRANSFER METHODS

After selecting the methods of feeding and orienting parts for transfer to assembly machines, the next major step in the conceptual development of an assembly system is to take the part from the feeder discharge point or escapement and place it in the work-holding fixture or product being assembled.

Fig. 5-29. This shuttle mechanism not only provides isolation from back pressure in the rail, but also modifies the attitude attained in the feeder to one of use in the assembly machine.

The construction of part transfer units varies immensely. However, certain key questions can be used to evaluate proposed design concepts. The important considerations are: reliability, safety of tooling and parts, and simplicity of maintenance. These three factors have an important bearing on machine justification and merit equal attention with initial capital expenditure. Among the evaluation criteria for parts transfer stations the following questions should be included:

• What form of motive power should be used for movement, timing, and control?
• In what direction does the part have to be moved?
• Is the transfer device to be used for final component orientation as well as for transfer?
• How is tooling safety to be provided?
• How much time will be required to clear insertion jams?
• Does the unit have enough inherent strength to operate at high speed over long periods without undue maintenance?
• What are the construction, installation, and operating costs?

• What effect will the design of the individual transfer stations have on the machine control system cost and on net machine production rates?

In evaluating proposed transfer design criteria associated with these questions, those customer personnel with procurement, program management or approval authority face a fundamental decision in choosing the assembly machine system, for at this point they fully share responsibility with the machine designer. In this chapter as elsewhere in the book, robots are treated as electromechanical transfer units.

As the starting point in the evaluation process, a logical question would be: "Is any transfer unit necessary at all?" Sometimes, there is no need for any transfer device, as Figure 5-30 shows. Here a cuplike cylindrical object sliding down the transfer rail stops against the indexing work dial. When the dial comes to rest, one part is allowed to slide into a fixture pocket and is then wiped away as the dial indexes for the next machine cycle.

This same type of wiping action can be used in feeding parts from a fixed or replaceable magazine-type feeding unit. The two necessary condi-

Fig. 5-30. Parts coming down the feeder tracks are swept away by the index motion without the use of a transfer device.

tions for success with this type of transfer arrangement are, first, that the parts must have such an outside shape that they can easily slide across each other without entanglement as the work fixture indexes and, second, that the work fixture, pallets, or dial must present such a surface to the end of the magazine or feeder track that the succeeding component parts are held in position during index without damage until the next work nest presents itself.

When parts are wiped from the bottom of a magazine-type feeder, uniform thickness of parts is often a critical requirement so that only one part is wiped away at a time and there is no tendency for subsequent parts to jam. This is often a major design challenge.

Figure 5-31 illustrates the use of detented rails, which are one degree higher in complexity than

the simple wiping action. In this case, the indexing portion of the machine does not hold the parts back in the feeder track. Instead, spring-loaded detents or shuttles, which can be mechanically or pneumatically operated, allow only one part at a time to fall by gravity or to be stripped away by the indexing portion of the machine. This is a concept that older readers first saw when snatching a brass ring on an amusement park merry-go-round.

The two transfer methods just described do not force the machine designer into any specific indexing mechanism, nor do they place any special demands on the control system. Any transfer arrangements more complex than these bear directly on the choice of a basic machine system approach.

There are three approaches to the transfer of parts that cannot be fed by the action of gravity

Fig. 5-31. Detents in rails hold back parts.

or weight as previously described. These methods are: mechanical units, fluid power actuated systems, or electromechanical devices. From a cost view, fluid power actuation is generally the least expensive, while mechanical transfer methods generally involve the highest cost. Robotic electromechanical systems are the most expensive; however, equipment cost alone is not a sufficient basis for judgment. The reliability of the transfer unit along with its ability to synchronize actions with other assembly operations is of equal importance. There is no hard and fast rule for evaluation. Each of the different types of transfer mechanism has its own area of application.

Single Motion Units

One of the simplest forms of transfer devices is the right-angle shuttle. In Figure 5-32 a pneumatic cylinder is used for a right angle transfer device. The transfer mechanism is simplicity itself but it depends on electrically or mechanically actuated valves to synchronize the insertion operation with the indexing function of the machine. In situations where a delay in valve actuation or cylinder operation could result in damage to tooling or parts if indexing starts prematurely, it is necessary to provide electrical interlocking devices that sense the location or proper position of such a fluid power actuated shuttle. The cost of the control equipment and the electrical interlocks must be balanced against

the higher cost of a mechanically actuated unit. However, cylinder actuated transfer devices have one advantage where rather long strokes are required. It is hard to build up long mechanical movements from the relatively short strokes of mechanical transfer devices. As a result, air cylinders have a definite advantage if a long stroke is necessary. When fluid power is used for actuation and is solenoid controlled, it is quite easy to integrate memory actuated control of motion with transfer device actuation. Completely mechanical transfer devices or gravity-type feeders have no provision for locking out parts transfer in the event of an assembly malfunction prior to that station.

Another common technique for transferring parts is shown in Figure 5-33. Often, the nature of the parts or the assembly operation requires that the part be dropped in from above. The least expensive way to perform this operation is by using a rigidly mounted overhanging transfer rail. Parts sliding down the discharge track from the part feeder are stopped against spring-loaded jaws. They are held in position until a plunger moving down from above cams the jaws aside and allows the part to be pushed through and inserted into the work-holding fixture or product assembly. Such units are used for screw or rivet transfer. The vertical plunger can be pneumatically operated or mechanically actuated from the basic indexing mechanism.

Although it is extremely simple in construction and very efficient, any rigidly mounted transfer device is difficult to clear in the event of a jam. Also, because it does not withdraw during the indexing portion of the machine cycle, this type of unit is more susceptible to damage during index than those more sophisticated transfer units that are withdrawn prior to work fixture index.

Many experienced in assembly equipment will not accept such rigidly cantilevered transfer units, but will insist on a more complex transfer device that is self-clearing and is less susceptible to damage. The self-clearing unit carries the spring-loaded jaws from some vertical moving tooling platform and raises them up out of the way of the indexing work fixture during the index cycle. In the raised position, the retaining jaws are level with the discharge track of the part

Fig. 5-32. An air actuated single motion shuttle escapes and transfers in a single motion.

feeder. One part is pushed by pressure, gravity, or a feed pawl from the feeder track into the transfer device. When the index is completed, the transfer unit is carried downward until the mechanism stops against the part piece or work fixture. At that point, the jaw mounting plate begins to compensate while the tooling platform continues downward. A center plunger pushes the part to be transferred through the detented jaws into the assembly. The vertical compensation of the jaw mounting plate is accomplished in two ways. This jaw is either spring loaded to permit over-travel or is fastened to one tooling plate of machines with double moving vertical tooling plates. The spring compensating or air compensating type is the more common.

Pick-and-Place Units

Up to this point, attention has been focused on transfer devices that move the part to be inserted in a single direction. Frequently, however, a pre-oriented part must be picked up from one position and carried over to another position for insertion. Devices that perform this type of transfer are usually called "pick-and-place" units. A simple pick-and-place unit that combines both mechanical and pneumatic actuation is shown in Figure 5-34. This type of unit picks up the part while the tooling platform is in its uppermost position during the machine index. The part is then carried down as the vertical tooling platform descends. This overlapping of the transfer function with the machine index cycle permits more rapid system operation.

A more common pick-and-place arrangement is the square motion transfer unit. In this unit the pick-up mechanism drops down vertically, lifts a part from the escapement, carries it forward over the indexing fixture, and then drops the part down into place.

Figure 5-35 shows a unit which is mechanically actuated from a main central camshaft. Two cam faces control both the vertical and horizontal motion. Since both cams are fastened to the same camshaft, coordination of these two movements is ensured. Moreover, direct coupling of this device to the main chassis camshaft coordinates the insertion activity with the indexing function of the machine.

Fig. 5-33. A single motion transfer unit combining a shuttle escapement and vertical motion of a plunger coming from vertical travel on the machine slide.

In addition to these mechanical units, electro-mechanical devices recently have been made available in which the stroke is controlled by means of limit switches. Another approach by other manufacturers of integral assembly machines offers standardized cam-controlled units directly connected to the center column of the assembly machine.

It is much more difficult to obtain a square motion than an arcing swing through a vertical plane. Since parts often have to be inserted through one another, a pure arc motion in a vertical plane could cause interference of one part with another.

Other pick-and-place units which pivot from one end have been developed. Figure 5-36 shows a pivoting type of unit in which the pivot axis is the centerline of the member that also serves to lift and lower the part piece. Such units experience great difficulty in maintaining locational accuracy. The slightest wear in pivot shaft keys, gearing, or linkage is magnified rapidly.

When parts are swung through an arc after being lifted from the feeder, they lose their previous orientation unless a compensating device is provided. If a pantograph mechanism can be installed inside the swinging arm, the escapement orientation of the part can be retained through

Fig. 5-34. A vertical slide operates this "lost motion" transfer device to carry the part from the rail into the fixture. Vertical movement of the head holds back parts still in the transfer rail.

the movement of the arm. In other instances, however, it may be desired not only to move the part but also to turn it radially during the transfer function. This rotational type of unit can be adapted easily to such actions.

Pick-and-place motions are used to transfer parts from feeders to work nests and additionally to move a subassembly in one attitude to another attitude. This may occur in the same fixture nest or another nest.

Some of the various commonly used transfer motions are illustrated in Figure 5-37. When feeding from an escapement or transfer track, the motion may be to follow a path similar to an inverted U as illustrated or be one in which the part is pulled from detents horizontally before being placed vertically down into the work fixture. For many reasons one might want to rotate the part radially during the transfer operation or roll it over to expose different faces prior to insertion.

Beyond these three simple motions, ingenuity can add to the versatility of rectangular coordinate transfer devices. By placing a simple lobe in the cam, a part during transfer can be dipped into flux, adhesives, or ink immediately prior to insertion. By placing two heads on a transfer device, a part can be picked up by the outboard head and placed in a simple date stamping, laser drilling, or similar secondary operation station while the inboard pick-up head will remove a previously processed component and insert it in the assembly. This technique permits more complex secondary operations on component parts immediately prior to insertion which require longer time periods than simple dipping or immersion, but this time usually cannot exceed more than 60% of the cycle time on the machine.

When transferring parts from one position in a work-holding nest to another or replacing the same part with a modified orientation into the same nest, a different set of conditions exist. Time becomes more critical. In taking a part from a transfer rail and inserting it into a fixture nest, the machine designer literally has the whole machine system cycle time available. The part can be picked up during index, positioned during index interlock, and inserted as soon as dwell is established.

The removal of a part and reinsertion into the same workholding pallet, however, can only be

Fig. 5-35. A modular rectangular coordinate transfer unit.

done during established fixture dwell time, which is often less than 50% of total machine cycle time. Some operations can be done in this short period, while others cannot. A longer time period is available if one can pick up a part during one dwell period, perform the necessary work such as rotation and replace it during the next dwell, but this is only possible if the part is inserted into the next fixture, since a fixture index has occurred during the machine cycle. This means that the part cannot be placed into the same location, since there is a part still in the next fixture that has not yet been removed. Reinsertion done after index must be done in another location or nest in the next fixture. Overlooking this matter at time of system design has caused many a red face!

Air Transfer of Parts

Compressed air can be used to power venturi units for transfer purposes. Such venturi units can be fabricated simply; however, inexpensive commercial units are available. In Figure 5-38 air

flowing into the side of venturi element develops a vacuum in the pick-off tube at the bottom of the unit. Parts oriented in a feeder are moved, one at a time, underneath the pick-off tube by a cylinder-actuated shuttle. The parts are then pulled through the venturi unit by the generated suction and finally are blown into the assembly.

Such units, when used for vertical transfer, tend to turn a part through an arc of 180°. The characteristic is sometimes a valuable feature, especially when parts can be fed more easily in an attitude other than that desired at the point of insertion. It is almost always wise to go along with the natural orientation tendency of the part feeder, and then turn selected oriented parts over in a feeder track or transfer device, rather than to try and force parts into proper position in the feeder.

In some instances, the transfer device also can be used to aid in the orientation of parts as shown in Figure 5-39. Here, the parts handled are pins with one round and one flat end. The pins are

Fig. 5-36. A pivoting-type transfer unit, once popular, which experiences severe difficulty in holding angular location.

easy to feed, end to end, but are difficult to select with round end forward for insertion. Since the machine is required to handle various pin lengths, it was decided that both orientation and transfer functions should be performed in the transfer unit itself. Operation of this unit is controlled by small air actuated shuttles. Often, such small shuttles in the transfer rail or tube are necessary to allow only one part at a time to enter the transfer device without jamming subsequent parts.

Transfer units often may be called on to perform operations other than that of inserting a part. Sometimes, it is necessary for the transfer device to properly orient parts fed in previous stations before inserting its own part. In other instances, the transfer device may be required to

perform a fastening operation as well as a transfer function.

Transfer Unit Selection

The motion performed by the transfer unit is usually settled by the shape of the part, the plane in which the part must approach the workholding fixture, and the position in which the component is oriented by the part feeder. The basic question in machine design becomes one of how to drive and control the transfer unit.

The completely mechanical transfer unit is the most dependable. The pneumatically operated unit can be the simplest in design, but its control and maintenance becomes more complex. An electromechanical unit must be rugged enough to withstand the demands placed on its relay or

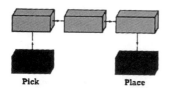

Pick   Place

If parts are feedable in the proper orientation the RPP performs simple pick and place motions in a single plane.

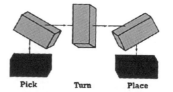

Pick   Turn   Place

Turning the plane of a part during transfer often leads to better machine layout, and better manual/visual access.

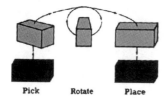

Pick   Rotate   Place

Rotation of the parts around their axis or center line may be required during transfer.

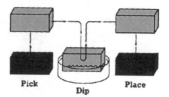

Pick   Dip   Place

Secondary operations such as lubrication, date or serial coding, imprinting or magnetizing can be done between feeding and insertion. Parts can be "dipped" as part of the primary motion or with specific dwell periods by a dual head adapter.

When it is necessary to expose other sides of a part for assembly, the RPP unit (or its sister RPU) can rotate any specified number of degrees and replace in the same or adjacent fixture nest.

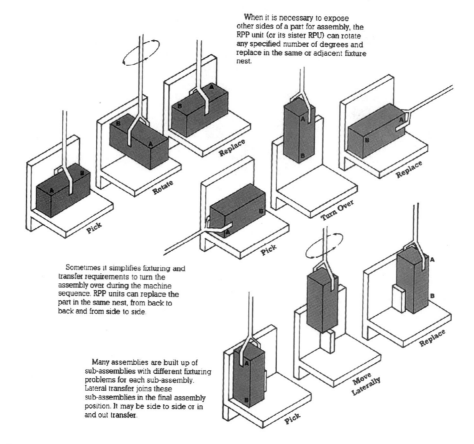

Sometimes it simplifies fixturing and transfer requirements to turn the assembly over during the machine sequence. RPP units can replace the part in the same nest, from back to back and from side to side.

Many assemblies are built up of sub-assemblies with different fixturing problems for each sub-assembly. Lateral transfer joins these sub-assemblies in the final assembly position. It may be side to side or in and out transfer.

Fig. 5-37. Various transfer unit motions found in transfer of parts to fixtures, or during assembly, from one fixture nest to another.

Fig. 5-38. Vacuum pick-up heads utilize compressed air flowing through a *venturi* unit which eliminates the need for vacuum pumps.

switching system by repeated high-frequency operations.

Before judgment is passed on the type of transfer unit to be selected, the type of machine control must be established. If memory type control systems are used, electropneumatic controls are the simplest to lock out for any given cycle of the machine. Many assembly system builders have standardized around a combination mechanical and pneumatic system in which advance of the tooling is pneumatically actuated while withdrawal is positive and mechanical. In this way safety to tooling is ensured and the pneumatic advance can be overridden by the control system of the machine.

When a simple instantaneous control is favored, rather than a memory system, a purely mechanical approach can be utilized with spring loading of the forward motion and positive mechanical withdrawal.

It would be presumptuous to state that any one design is superior to another, for the transfer station cannot be judged alone. It must be viewed in its relation to the total machine system. Each builder of standardized modular assembly equipment uses certain basic transfer designs, which may vary from job to job in minor details. If one chooses such a builder, one also chooses to accept the builder's transfer units. It is a case of love me, love my baby. But each of the transfer designs must be compatible with the basic indexing mechanism and control system used.

It would also be presumptive to pretend that this discussion covers all types of transfer units. However, it does cover the basic motions that are utilized on the majority of transfer functions.

For those personnel involved in the procurement of assembly equipment, the questions listed at the beginning of this chapter provide a checklist for guidance in deciding on one of several competitive systems. The important point to be kept in mind is that individual transfer stations are not isolated operational units even in power-and-free transfer systems but must be compatible with the entire system. A sound assembly machine design often incorporates the best features of all systems to ensure a satisfactory solution.

SUMMARY

Perhaps no industrial equipment can command so many human emotions as the performance of an assembly machine. Rage, frustration, and anger are common, but so is the unadulterated joy of watching a well-designed and well-executed assembly machine. The ingenuity that goes into successful part feeders, escapements, and transfer devices is a pleasure to behold. But just as one cannot teach what makes one artist great and another mediocre, one cannot expect to find in this book the creative tool design solutions to unique problems. If Edison did truly say that genius is composed of 1% creation and 99% perspiration, we can only add that this book is intended to significantly reduce the amount of perspiration. Success in automatic assembly design and operation comes from an awareness and sensitivity to a multitude of small details; and by avoiding the tendency to assume that any one person will have sufficient knowledge to create a perfect machine without any misfortune along the path.

Perhaps no more disturbing problem can occur in the development of an assembly system

Fig. 5-39. Pins of varying lengths with one round end are oriented and transferred in the same device.

than the inability of a tool designer or feeder builder to admit that the parts feeding concept they have chosen for any given function is not adequate to the job. The first instinct when trouble occurs is to feel that the station is bad in execution, rather than concept. That is not only possible, but often the only problem. Debugging is the refinement of the execution of sound concepts. It takes a good deal of humility and maturity, however, to admit that in some instances the very concept is inadequate for the job at hand. Everyone must be prepared in the debugging stage to ac-

cept that a certain percentage of the tool design decisions may be in error. One must avoid the pasting of bandages over incurable cancer, of interminable addition of air jets and spring wire to a marginal station.

Delay in determining if feeding and transfer stations are the result of poor adjustment, poor manufacturing execution of a sound design or in fact is a poor design can undermine customer confidence when such confidence is vital to program completion.

Chapter 6
Integrating Fabrication and Packaging Operations with Assembly

INTRODUCTION

In those companies where fabrication is done departmentally or a significant part of the components are purchased externally, the very high cost of handling, transporting, and reorienting fabricated components so that they might be assembled, together with the damage that such handling causes, mandates a serious look at direct coupling of fabrication and assembly operations. Today, many companies have restructured plants to "work cell" configuration to address this problem. This chapter does not address this issue. Instead it addresses fabrication and packaging directly on the assembly system.

In many cases, direct coupling of fabricating equipment to an assembly system is not economically feasible and in other cases is technically impractical. There do, however, remain several areas where direct coupling of fabrication and assembly is the only viable way of automatically handling certain types of parts.

There are many factors to be weighed. Among them: Is the last fabricating operation prior to assembly a primary or a secondary operation? Many secondary operations are completely compatible with and can easily be combined with as-sembly operations, particularly if no chip making is involved.

Does the fabrication work involve coolants, parting waxes or lubricants? If this material remains on the component, is it compatible with the subsequent assembly operations? This is usually a negative factor.

Are trimming, deflashing or degating operations required prior to assembly? Should this work be integrated with the assembly operations to ensure product quality?

Does the primary operation require such long cycle times that multiple cavity dies, molds, or other means of multiple fabrication are necessary to balance assembly requirements? Are several fabricating machines necessary to produce sufficient parts for the assembly line?

Direct coupling of component parts fabrication onto the assembly system is very practical in four areas. In fact, it may be the only feasible way to feed and insert certain types of components. These areas are secondary machining or forming operations, press operations on fragile or extremely thin parts or those parts that would tangle readily, spring forming operations and insertion of parts that are best fed or made from coil form such as tubing or wire.

Coupling may occur by total fabrication of the

component at the assembly machinery or by doing final forming and cutoff operations to preformed parts coming to the machine in reels or bandoliers.

It should be noted that some fabricating operations often must occur after some preliminary assembly work has been completed. Such integration should not impede further automatic assembly operations.

For many reasons, controlled ejection of the products that have been assembled may be required for successful integration of the automatically assembled product into the final manufacturing line. It is mentioned elsewhere that many otherwise successful assembly machines have been downgraded to marginal operations because of failure to consider the nature and accessibility of the ejection station as it relates to product flow through the plant to final assembly and/or packaging.

In many cases assemblies are fragile. They may require controlled isolation from other assemblies. They may require additional cure time or bonding before shipment or final assembly. They may have to proceed to lengthy automatic testing operations incompatible with assembly line speeds.

Ejected assemblies may be fed as a single component to other subsequent assembly machinery or directly to packaging machinery. Retention of established orientation is strongly indicated here, but it must be done in a way that is economically practical. Conveyors, accumulators or other forms of buffer storage, and pallets must be viewed in the light of their economic feasibility (as well as the reduction of possible product damage and its cost) when compared to bulk ejection and subsequent automatic feeding.

In this chapter we will tend to ignore component parts with substantial physical integrity having a solid dense configuration and few, if any, fragile protrusions or projections. Such parts, particularly those which require multiple fabricating operations (for example, a slotted head screw), will best be fed to the assembly system from bulk storage through the use of conventional parts feeders. This chapter will concentrate on component parts that are fragile, that have slight asymmetry, that can be damaged by surface abrasions and, particularly, those parts that

would tangle if brought to the assembly machine in unoriented bulk condition, or where appearance and/or cleanliness is critical.

Before proceeding it would be worthwhile to mention the specific area of electronic assembly. At this point in time, however, most automation in electronic assembly consists of the insertion of leaded devices (such as resistors, capacitors, and diodes), surface mount components and integrated-circuit devices with external projections (such as those housed in DIP packages). The problems of feeding such parts are essentially those of specialized escapements and transfer devices since the product usually has been placed in bandoliers, film carrier reels, or in magazines at time of manufacture. These specialized placement devices usually include some secondary operations such as lead or terminal forming, and part insertion, cutoff, and crimping. They do include some of the characteristics discussed in this chapter.

Most experts, however, would exclude electronic or circuit board assembly from conventional automatic assembly. The growing use of leadless electronic components, together with standardized modes of component packaging, will further remove direct integration of electronic components fabrication from the traditional assembly process.

Lastly, this chapter will emphasize the growing importance of buffer storage as vertically integrated plants are increasingly arranged in work cell layouts rather than layouts arranged departmentally by specific technologies.

It is necessary to point out here, as was noted in the chapter on product design, that successful integration of part fabrication and assembly will require complete teamwork and cooperation among Purchasing, Product Design. and Manufacturing Engineering. In theory, this is what simultaneous engineering is all about. In practice, it may be something less.

PRESS OPERATIONS

Die-casting machines, plastic molding presses, heading machines, and conventional presses produce component parts through the use of pressure on material. In earlier times, the vast majority of products produced on such primary fabrication equipment required some secondary operations, such as drilling, reaming, undercutting, grooving, and tapping, particularly where fasteners were involved. Present-day product designers trained in net-shape manufacturing, aware of the costs and quality issues of secondary operations and having a wide array of new fasteners and connectors, have designed out of their components much of the necessity for secondary operations.

This would seem to enhance the possibility of integration of primary fabrication and assembly operations. A closer view of the first three types of pressure-forming equipment begins to weaken this possibility.

Headers are extremely high-speed machines capable of producing parts far in excess of the average assembly machine's ability to consume the header's production. The use of lubricants, the high rates of production, and the high cost of heading machinery are strong factors against the feasibility of dedicating a header to feed one automatic assembly machine.

The production of parts in die-casting machines is usually high, not because of their cyclic speeds, but through the use of multiple-cavity molds. Single-cavity molds would rarely keep up with production demands of high-speed assembly machines. Multiple-cavity molds, however, often place each cavity in a somewhat different radial position on the mold face to facilitate material flow through the sprues. When degated and trimmed, each component part has a different orientation. The nature of the degating operation and die-casting cycle times usually makes direct coupling of die-casting machines to assembly machinery impractical.

Injection molding and blow molding machinery may seem more adaptable to such direct coupling, but the slow cyclic rate, the large number of cavities, and the degating process usually preclude this option. One notable exception would be where products from the molding machine cannot stand abrasion. For example, camera lenses molded by Polaroid are degated from the sprues directly into the assembly process. Dramatic changes in molding technology and mold designs, such as hot runner systems, allow for direct transfer to conveyors or magazines.

This wide use of molding sprues as a carrier magazine is a most interesting one. Consider a computer terminal or calculator keyboard. The individual entry buttons or keys are physically alike but with one significant exception, the printing or marking of each key top to identify the function of the key. If the entire keyboard matrix can be then molded in one shot with the correct imprinting and degated from the sprue directly into the assembly, a very real problem in part orientation is solved. The successful use of the sprue to integrate molding and assembly can be a very effective tool in the integration of molding and assembly. It does assume, however, that the mold designer and mold maker recognize the dual role of the sprue and can position the mold cavities on a grid pattern and in a radial orientation consistent with their final insertion pattern, as illustrated in Figure 6-1. This may be in conflict with optimal press capacity.

Very close coordination will be required between the mold designer and assembly machine designer to ensure that degating is efficient and that the sprue marks are not located in the insertion area. The advantages of such integration must be weighed against several factors. Where

Fig. 6-1. Individual calculator buttons are molded in groups exactly as they will be used in the assembly. The entire button molding is degated as part of the assembly process.

there is broad product mix, such as is found on scientific calculators, the cost of dedicated molds of the proper matrix may be prohibitive when batch sizes or product volumes are small. A blemish on one key may mean discarding of the whole sprue. Shearing away the sprue at insertion may mean contamination of the assembly or present cosmetic problems. From a production standpoint, the material used in the sprues may be reusable if not contaminated. The salvageability of the sprue material can be an economic consideration when the cost of resin fluctuates steeply.

With the exception of using the sprue as a carrier or part magazine (and that is limited to where the parts on the sprue may be captured or used in the existing orientation) there usually is no cost-effective way to simply couple molding machines with assembly machinery without the use of part feeders, or by the increasingly important use of buffer storage conveyors and storage magazines. The decision for conveyor or magazine storage must predate mold and extractor design.

Complete Fabrication of Stamped Parts

Parts that are stamped out of their plastic, metal, rubber, or paper material may well lend themselves to complete fabrication directly on the assembly system. Certain fabricated materials such as styrofoam, rubber, mylar, and paper are very difficult to feed from a disoriented bulk condition. They become sticky and often acquire static charges. They are sensitive to heat and humidity as well as static charges. Blanking parts into magazines or piles and then attempting to coin exchange them or vacuum the top (or bottom) piece off the pile can turn out to be an exercise in frustration, as shown in Figure 6-2.

Fabricating such parts in a small press mounted on or adjacent to the assembly machine is strongly indicated when one or more of the following characteristics are present:

- The parts are less than 0.010 in (0.254 mm) thick.
- The nature of the material is subject to change because of atmospheric conditions (e.g., temperature, humidity, or electrical change).
- Parts tend to adhere to one another.

Fig. 6-2. Styrofoam material used for gaskets in bottle caps clings and resists feeding when prepunched. It is best fabricated directly on the assembly machinery.

- Gauge thickness may vary widely as a percentage of overall thickness.

The machine designer when opting to incorporate direct fabrication on the assembly machine has several important considerations. All of these in some way relate to machine productivity. Before proceeding too far the designer must ascertain in what form the material is available for fabrication: sheets, coils, strips? What is the relative cost of different material forms? What percentage of the material is left in the web as contrasted to the material in the individual components? Is there any salvage value to the remaining material? How often will reels of stock have to be changed?

Usually the machine and product designers will run into resistance from Cost Accounting and Purchasing at this point. High-volume stamped parts such as gaskets or washers used in manual assembly are often stamped out of fairly wide material using multiple-cavity dies so that the greatest part of the material is used for the product.

Fig. 6-3. A simple blanking press pushes stamped part pieces directly into workholding fixtures.

Ingenious die layout can produce a web that may be rerun through another die, reducing material cost dramatically. Small washers can be blanked out of the web from blanking larger washers. One component part may be run through large presses at high speed, and the press can then be changed over to other work.

A decision to manufacture such a part on the assembly machine will usually mean a much poorer ratio of material usage to remaining web. This increment of increased material cost must be balanced against the assembly savings by removing the necessity of manual assembly. There really are few options when a component is thin, fragile, or subject to atmospheric conditions. If it cannot be fed from bulk automatically, it must be fed by hand or blanked from strip stock and then transferred directly to the assembly. We assume that product design has found no way to eliminate such a component piece.

For most assembly machine operations and from a machine design standpoint it would be preferable to have the material coming to the machine in coil form from which a single part can be blanked.

The designer is now free to concentrate on certain key assembly considerations. Should the press unit blank the part directly down into the assembly as the punch travels through the die? This is a simple answer and often the right one (Figure 6-3). The concern of course is that the die must be rigidly cantilevered over the indexing fixture. Any jams may cause extensive downtime and possible fixture damage.

A second design choice is to invert the punch and die and blank a part upward into a reciprocating transfer device that will carry the part out into the assembly fixture. This approach creates a much safer operating condition.

A variant on this theme may be to incorporate a vacuum into a reciprocating punch allowing it to both blank and then carry the part to the assembly fixture.

Other key design features must provide for excellent alignment of punch and die, ease of removal of punch and die for maintenance, and rapid renewal of material when one coil of material runs out.

Alignment and press rigidity will reduce the frequency of die maintenance, but in any event those choosing to fabricate on an assembly machine must be prepared to accept the downtime that die maintenance will cause the entire assembly system, unless there is a sound preventative maintenance program.

Paper and thin plastics such as mylar present greater problems when zero clearance dies are required. If not properly maintained, the punch will tend to pull the material through the die.

Any press station will need some provision for stock feeding into the die set. Occasionally when feeding material with elastic properties or with a tendency to resist being pushed into the die, it may prove necessary to pull as well as push the material through the die. When pulling the web through the die, material distortion may occur, resulting in distorted blanked parts coming from the die. If this occurs, it may be necessary to release the material to a free condition immediately before blanking.

Handling the web may prove a greater challenge then blanking the part. Since material flow to the machine press station will generally be perpendicular to the movement of the work holding fixture, the web will often be ejected from the press toward the machine. If there is no requirement to save the web it can often be chopped up into small pieces by the action of the die set. This simple answer can prove troublesome when static electrical charges make disposal of the scrap material difficult. Housekeeping problems are compounded if this scrap material tends to cling to the product. In this case rewinding the web on a take up spool may prove the only way to handle the problem, as illustrated in Figure 6-4. Because of material cost factors, it may prove necessary to wind the web so it can be run through a second or third time or more to blank out additional parts.

Fig. 6-4. In this press design, the punch travels from the die set to the transfer point in an inverted U motion.

Coil size will be an important judgment area. In general the largest possible reel size will be the most practical, since it will allow longer runs without changing coils of stock. This is truly important when only one part is fabricated from coil stock on a machine. When several coils of stock are used on a single machine, reel size is an even more significant factor. Every time a coil is changed the entire machine is down. Machine operators are forced into real dilemmas when one coil has run out and other coils are nearly exhausted. Should they change all the coils at once discarding whatever material remains, or should they change each reel when it is empty?

When it is necessary to rewind the web, particularly if it is to be reused, coil size may be determined by the operator's ability to load or remove full reels of stock.

So far we have been talking about relatively thin parts and fairly simple die sets rigidly fastened to or adjacent to the machine.

Completion of Preformed Parts

When simple blanking operations are sufficient to produce a part, required press tonnage is modest. Highly efficient over-center toggle presses or fluid-power presses can readily be integrated into the machine design.

When component dies are required to produce deep drawing operations or for complex forming operations, higher tonnages may be required. Die sets become larger. Die maintenance becomes more extended. Die costs may preclude stand-by dies for quick replacement. Required press size may not permit good access to the assembly fixture.

The machine designer under these conditions should back away from direct physical integration of the press to the assembly machine and consider doing the primary processing of the components in a conventional press operation at optimum speeds, while leaving the processed parts attached to the web or carrier strip. This then is rewound and carried to the assembly line where it is unwound and individual parts are cut off from the web on the assembly machine. This approach has been used successfully in some industries for years. AMP electrical terminals are a typical successful approach to the type of problem. Examples are shown in Figures 6-5A and B.

The benefits are obvious. Isolation of die maintenance from assembly machine downtime is most important. The assumption is, however, that the parts can remain in the strip without damage or distortion when rewound and stored for periods of time.

Paper or plastic interleaving may have to be introduced in the rewinding process after the primary forming operation and then removed during the assembly phase to prevent tangling of parts retained in the web with previously wound parts.

Forming Operations on Subassemblies

While carrying partially completed parts in the web or strip is widely utilized, a very closely allied opportunity is rarely used. That is the secondary or final forming of discrete parts as part of the assembly operation. We are not talking at this point about press swaging-type operations used to join parts or to complete an assembly.

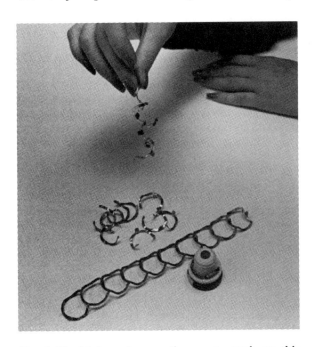

Fig. 6-5A. U-shaped mounting parts such as this would tangle as discrete parts, but the die size and tonnage requirements preclude direct mounting of the press on the assembly machine. Preformed parts are carried in the strip where they are cut off for transfer and insertion.

This is covered in some detail in the chapter on joining. What is intended here is to describe how final press forming operations integrated with the assembly machine may facilitate the assembly of difficult discrete components.

The problem one faces in complex or even simple assemblies is that *some discrete components cannot be automatically fed, oriented, transferred, or inserted on a reliable basis in their final shape or form.* They may become tangled, interlocked, or distorted in handling. They may lack sufficient integral strength to be inserted or transferred in their final form without damage. *Yet at some intermediate stage in their fabrication they may have a form or shape that will allow reliable automatic handling.* It may be that, in some instances, stiffeners or webs could be left in place to allow the part to be fed or handled and successfully inserted. Following feeding and insertion the part may then be formed to its final shape, or stiffeners or webs sheared away to leave a functional component. (See Figure 6-6.)

This same approach can be used to produce an assembly when the use of multiple fragile wire forms or stampings would jeopardize efficiency. By feeding a smaller number of wire forms or stampings, and properly joining them, a blanking operation could then remove connecting material to provide a functionally correct assembly. This approach will, of course, require that product design, fabrication, and assembly work hand in glove. This matter is mentioned in detail in Chapter 2 on product design.

In order to form (or separate) parts that are fed, inserted, or joined in an incomplete configuration, at least one or more of three conditions must be met. First of all there must be access for the final forming or blanking tool to function. Second, there must be provision for the tools to withdraw after forming or shearing. Third, there must be provision for the removal of any material separated by such secondary operations.

There is often an assumption that a component being formed to final shape can be formed over another component of the assembly. This may or may not be true. One or two simplistic tests are not sufficient to commit the building of a major assembly machine system. A great deal of certitude should be required before assuming that an-

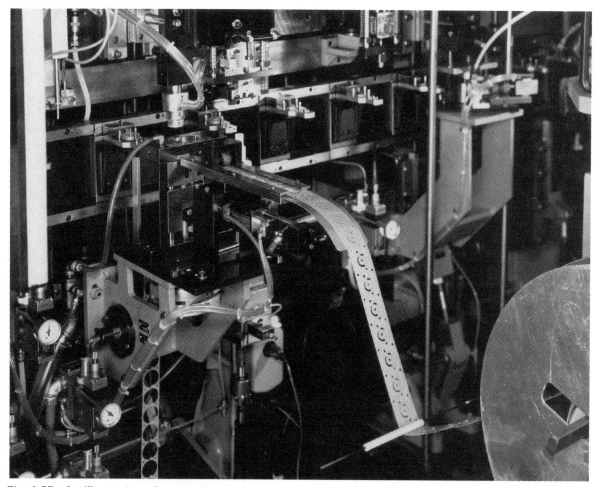

Fig. 6-5B. An illustration of a partially blanked preform ready for blanking into the final assembly.

other component of the assembly will act as a bending mandrel without itself being damaged.

If tests or design prove that using another component as a forming tool is not practical, there is a tendency to incorporate such forming mandrels into the workholding fixture. Such a move should be discouraged. In other places we have discussed the problems that occur in fixture maintenance and locational accuracy when riveting mandrels, lower electrodes, or similar items are incorporated into the fixture. Accidental damage to one fixture may go unnoticed until defects are found in the product being assembled.

The use of forming or riveting mandrels, or welding electrodes as part of the fixture, particularly when they have to be actuated, is an easy

way out for the machine designer. Operating problems will rarely be noticed during short acceptance runs or preliminary installation. In real production operation, however, the problems of excessive fixture wear or damage become increasingly severe. There will be occasions where this approach is the only viable alternative. If this is so, then one must accept it only with design provision for easy replacement of fixture components subject to wear or damage.

The ideal solution to forming components would be to have both press and forming tools (or punch and die) retract at each index. The design work on such stations can be simplified if access for forming tools can be made through the assembly by inverted tooling. (See Figure 6-7.)

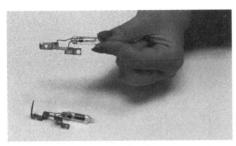

Fig. 6-6. Rather than feed two terminals to be welded to individual wire leads, a common stamping was fed and welded to both leads. This stamping was parted in a subsequent assembly process to provide a functional product.

SPRING WINDING OPERATIONS

One of the most difficult component parts to handle in assembly mechanization is any form of wire spring. Advances in feeder concepts allow routine feeding of some springs formerly considered poor candidates for mechanization. Certain

Fig. 6-7. A double knuckle action press moves forming tools, both punch and die, into position from the inverted position before forming pressure is applied from above.

spring forms, however, particularly open ended springs and conical spring shapes, interlock so much that feeding from bulk is not practical. Spring fabrication as an integral part of the assembly process should then be considered.

Springs come in many forms—cylindrical, conical, flat and dog-legged. They are expected to resist or produce compressive, tensile, and torsional forces. Flat springs often are susceptible to vibratory or magazine feeding techniques. Flat springs are made on presses. Any attempt to integrate their fabrication would follow along the lines of press operations described earlier in this chapter. Other spring forms are usually formed on spring formers, spring generators, or, more generically, spring winders. These are machines in which a winding mandrel is coordinated with wire feed, wire forming, and cutoff tools that are controlled by mechanically coordinated camshafts, gearing, and cams.

Spring winding is still very much an art, a trial and error approach that must be recognized if one attempts to integrate spring manufacturing with assembly.

Spring quality and consistency is not only determined by the skill of the winding machine setup man, but also by the characteristics and diameter or gauge of the material being formed. Simply changing a reel of wire stock may change spring characteristics. The implications are that any attempt to coordinate spring manufacturing directly with the assembly machine operation must accept the very real dangers of significant system downtime when changing coils of wire or when tools are worn. It may be necessary to have Quality Control qualify springs from each new roll of wire or after any tool replacement on the winders. In the meantime, the assembly machine will sit idle. Extremely small changes in the diameter of the wire from coil to coil can have significant effect on spring characteristics.

Spring Winders

Spring-forming machinery is designed for universal spring-making work within certain parameters, such as numbers of coils and certain wire gauges. They are mechanical in nature and are set up (by trial and error) to produce a specific spring form. They are also designed to operate at high speeds on continuous operations and are rarely sold for single-cycle operation or tooled for a specific application. Those purchasing a spring winder to be directly coupled to an assembly machine must be prepared to provide for single-cycle actuation usually through a clutch-motor installation, rather than through direct mechanical linkage to the assembly machine's camshafts. The ability to single cycle the winder independent of the assembly machine will be most useful in setting up or adjusting the winder. In mounting the winder on the machine or adjacent to the machine, one must provide for good manual access for adjustment of the forming and cutoff tools. (See Figures 6-8A and B.)

In providing for transfer of the spring, it must be remembered that springs exhibit unique characteristics at cut-off, so capture of any spring and feeding it to a carrier tube or track may require a great deal of experimentation. If the fabricated spring is to be transferred some distance, it may not pay to feed directly from the winder into the assembly since springs in transit (often carried in an air stream) may develop significant inertia and bounce on contact away from their proper loca-

tion in the assembly. If a venturi or suction unit is used to transfer the spring through a tube from the winder to the assembly fixture, it may be necessary to place a decelerating device in the transfer tube. This may be as simple as a slotted section of tubing to dissipate the vacuum or the introduction of a timed escapement holding one spring directly over the assembly fixture and releasing it to fall a short distance without significant inertia. It should be noted again, in developing high volume assembly machines, that the forces of gravity are subject to specific physical laws, and that a free body will not fall a greater distance in a given time than those laws dictate. Many designers completely overestimate the forces of gravity on a free falling body.

There are many successful applications in which spring winders are directly coupled to assembly machinery. In opting for direct coupling, it usually is assumed that the spring may be used in the assembly without secondary forming as a separate operation. They also generally assume that there is no requirement for spring tempering operations as an isolated operation after winding and before assembly. It may be possible to temper cylindrical springs by passing them through an induction coil as part of the feeding process.

Secondary Forming Requirements

Many dog-legged torsional springs and some tensile springs may require secondary forming operations before going into an assembly. There are often some bends required to engage or locate the spring in the assembly that cannot be made as part of the primary forming operation. When there is an absolute requirement for a secondary spring forming operation, the machine designer may find it necessary to feed the spring from a vibratory feeder and hence integration is not necessary. The designer may be able to capture a spring on ejection from the secondary operations and load it onto a mandrel or in a magazine for subsequent transfer or may be able to transfer it directly from the secondary operation to the assembly.

Additionally, the secondary operation might be performed on the assembly machine itself in a separate fixture pocket. Springs fabricated in a

Fig. 6-8A. A commercial spring winder designed to fabricate springs on demand and feed them directly into the assembly machine.

Fig. 6-8B. When tempering of springs is required, a tempering oven may be placed between the spring winder and the machine.

primary operation could be fed directly to a separate nest on the assembly pallet and there receive their secondary forming before being transferred into the assembly itself.

Tempering Operations

In the winding of springs, stresses are induced which will in time relieve themselves causing some changes in spring characteristics. Product design specifications may demand that these stresses introduced by manufacturing be relieved thermally by bringing the springs to a known temperature for a specific period of time without destroying the metallurgical characteristics of the spring wire. If the spring can be fed from bulk condition automatically, the tempering operation presents no burden to the assembly operator. If, however, the spring must be fed directly from the winder to the assembly fixture because of its tendency to tangle or interlock, the requirement to temper the spring may pose additional problems. A simple change in spring material may eliminate

the problem. Going from conventional music wire to stainless steel wire may prove sufficient.

Flexing the spring several times during assembly or functional testing may prove sufficient to relieve stresses. The actuation of a spring several times as part of the assembly process may do enough to relieve induced stresses. Subsequent stress relief will not significantly alter spring characteristics.

When product design is adamant that thermal stress relief will be required, there is another option available to the assembly machine designer: placing an induction coil on the spring winder through which the spring must pass.

It has been noted earlier that spring winding is an art. As in so many artistic endeavors, past practices are venerated and spring making is no exception. Although repeated tests have shown that there is no discernable difference between springs tempered instantly by induction energy and those slowly cooked in tempering ovens, it is hard to gain acceptance of induction tempering of springs. Thus we are led to buffer storage of springs.

Spring Storage

Since the ideal goal of all mechanized assembly is to reduce the direct labor content of assembly, direct coupling of spring winders to assembly machinery seems quite logical at first sight. In fact, however, it involves dedication of a universal fabricating machine to a specific task; operation of that machine at below optimal rates; and requires a rapid availability of underutilized skilled indirect labor at the machine site. Normally one skilled setup man could run a battery of spring winders.

When feeding springs from bulk is not practical, and when a long tempering operation is necessary, buffer storage in a controlled state will be required. There may be reasons to go to such interim storage between fabrication and assembly even where no tempering is required. For example, stringent quality control inspection of spring characteristics may be necessary between fabrication and assembly.

Attempts to package springs to facilitate assembly have tried various techniques including placement of springs in tubes, on paper trays, on sticky-back paper, and on strings. All of these attempts have been less than successful.

Development work in the automation of military fuse assembly led to a spring storage system that solves several problem areas. Parts can be fed directly to the storage magazines from the winder operating on a fully automatic basis. The same magazine system provides for easy transfer of parts from storage to the assembly system. Storage density is high. And lastly, the magazine plates can be loaded and then placed in tempering ovens to placate the most fastidious spring specifier.

The storage system is illustrated in Figure 6-9. It uses circular 30-in.-diameter (76-cm-diameter) aluminum plates in which a series of holes to contain the springs are placed in the form of an Archimedes spiral. These plates are placed on a carrier with freedom to slide in both the X and Y axis. The magazine plates are indexed under the transfer tubes from the winder and to the assembly machine by an inverted bullet nose pin with a U-shaped motion that causes the plate to rotate and at the same time adjust in the X and Y axis so that the spring pocket is always prop-

erly placed for pick up and transfer. Such a hole pattern can be cheaply produced in inexpensive aluminum tooling plates by the use of numerically controlled machines readily programmed to an Archimedes spiral pattern.

WIRE TERMINATION

In Chapter 2 on "Product Design for Automatic Assembly," the elimination of electrical lead wires was indicated as a most desirable goal. Unfortunately, many products suitable for automated assembly have a need for electrical connections. The assembly of wire harnesses in itself is a major industry. It is an area in which a great deal has been done to prepare individual wires for manual assembly but in which almost nothing has been done to connect terminated wires efficiently to other objects.

This difficulty is rapidly being resolved through major product innovation, increasing use of fiber optics, the advent of insulation displacement connectors specifically designed for automatic assembly, and the use of hard wire forms using air or nonconductive tunnels for insulation. These developments have greatly increased the possibility of automating the assembly of small motors, solenoids, and similar devices.

This is a logical place to mention, however, the automation of wire termination: the cutting, stripping, and termination of insulated wires, so that they can be utilized individually or in harnesses. Western Electric and Packard Electric were pioneers in this effort because of their large in-house requirements.

While early efforts merely stripped insulation, the necessity of terminating wires to ensure good electrical continuity and ease of maintenance led to the development of a wire termination industry manufacturing terminals generically, if often incorrectly, referred to as "AMP" connectors. To facilitate automation, these terminals were designed to be manufactured not as individual pieces but in some strip or bandolier configuration from which they could be automatically cut or removed and clinched in a swaging operation to previously stripped wires at very high rates of speed. The major terminal makers developed their own equipment for lease to major cus-

Fig. 6-9. Circular storage plates with spring pockets drilled in an Archimedes spiral provide dense storage and easy transfer of parts, allow for tempering, and protect fragile springs from damage.

tomers. An independent wire stripping machine industry has developed as well, allowing users to purchase terminations from multiple sources.

Successful automated wire termination will depend on the quality of the terminals used but more importantly on the attachment of the insulation to the wire core, the insulation diameter consistency, and the concentricity (or uniform wall thickness) of the insulation.

Single-end and double-end termination of individual wires is commonplace, and fluxing and soldering of terminals to wire successfully are done with standard commercial machinery. Automation of two-wire and three-wire conductors is commonplace, and there is a growing capacity in termination and testing of electronic conductors such as ribbon wire. Large programmable machines are under development.

What is not being done is to integrate wire termination as part of automatic assembly of electromechanical devices or to prepare large harnesses. Major research funded by civilian sources such as Packard Electric and the U. S. Army project, "Mechanized Assembly of Harnesses" (Report No. OR-15-415) have not led to economically feasible automated harness assembly systems.

In brief while wire termination per se is widely automated, it is not usually capable of easy direct integration into automated assembly lines. Most advances in assembly automation will come from producing products using a minimum of insulated wire components and having bayonet or similar external terminals, for later attachment to wire leads.

SECONDARY MACHINING

Several major builders of assembly machinery started out as metalcutting machine manufacturers who were asked by customers to add mechanized assembly operations to what was essentially a chip making machine. Early efforts included the insertion of screws, pins, and bushings or sleeves to a part being machined. Further demands for improved productivity led to automatic feeding of the part to be machined into the workholding fixture on the secondary operation machine.

Conventional Machining

The tail now wags the dog and builders of automatic assembly machinery are often asked if they can include secondary machining operations as part of the assembly process, thus integrating fabrication with assembly. The answer is a very qualified yes. Assembly machines have little or no provision for ridding themselves of chips, coolants, or lubricants. Assembly fixtures usually are not designed to withstand machining torques. Contamination of the assembly being processed is easy and cleaning it by air jet or vacuum difficult.

For the most part, therefore, secondary machining using rotating tools is not a desirable situation on the assembly machine. Broaching of slots or punching of holes is much more adaptable subject to the considerations mentioned earlier in this chapter. Increasing use of a new technology, laser machining, is very compatible with automatic assembly.

Laser Machining and Marking

It is the contamination from removed material and coolants or lubricants that is the basic cause for most assembly machine builder's reluctance to incorporate secondary machining into an assembly system. The necessity of staging parts to resist machining pressures or torques is a second factor. These problems do not occur as often with the use of laser machining or marking. The material removed is in a gaseous state and can (and must be) vacuumed away. There are no real pressures induced on the workpiece in material removal or marking operations, thus simplifying component design and fixture construction.

The machine designer has several considerations when using lasers to machine or mark assemblies. SAFETY is paramount. Most are aware of the necessity to avoid the laser beams during operation. This "safety" attitude must also be built into the provision for laser position adjustment, focusing, and calibration, as well as obligatory use of special guards and goggles.

Not so readily apparent is a real concern about the disposition of material in the gaseous state generated by laser machinery. Means for the immediate removal of this material should be designed into the machine.

From an operational standpoint optical alignment of lasers is critical. The problem is to find a stable mounting surface for the laser optical system. Several builders have found the use of granite surface plates, often used for shop inspection or layout purpose, as ideal mounting beds.

When selecting lasers for incorporation into assembly equipment, the machine designer has to be concerned with the required system cyclic speeds and total cycle time possible with the power source quoted. CO_2 lasers are quite compatible with automated assembly; other laser power sources may not be so compatible.

It should be noted whenever discussing the possible use of any emerging technology such as lasers, that the builders of new technology may not have any significant production experience with the new equipment, nor made realistic provision for its integration on an assembly system. Additionally, because of the sophistication of the equipment, the inventors (often scientists or academics) may have little experience with or sympathy for the industrial environment in which it must operate. The assembly machine builders must consider not only the theoretical capability, but the endurance of any emerging technology in the proposed operating environment.

Laser marking of products is a particularly interesting development for products requiring data coding or other identification marks, particularly where the material of the products being assembled, the design of the assembly, or the environment in which the product is being used will allow neither pressure generated identification (stamping) nor ink marking.

EJECTING TO PACKAGING OR IN-PROCESS STORAGE

Integration of assembly operations with fabricating operation assumes an integration with those manufacturing operations preceding or concurrent with assembly, but assembly is usually not the last step in the manufacturing process. Testing, packaging, or integration of the product just assembled within a larger assembly are normal events occurring subsequent to initial assembly. Ejection of a staged assembly, one having a known attitude and spatial orientation, should be done in the light of any possible downstream integration. In many cases simple ejection to bulk storage bins will prove the most cost efficient approach; in other situations controlled ejection to some form of storage will be indicated.

It has been mentioned elsewhere that one of the most common errors in assembly machine design has been the tendency to look at the ejection station as an insignificant part of the systems design. Room for storage bins, tote pans, or conveyors shrinks as the more active stations grow in scope and complexity, all tending to occupy more and more of the space originally intended for containers, conveyors or pallets to receive completed assemblies. Sometimes access to the ejection station becomes so restricted as to force the use of a conveyor to carry completed parts away from the machine to a more open area. Such conveyors tend to impede the machine attendant's easy access to stations on the other side of the conveyor, thus increasing downtime in the event of jams or work flow stoppages. (See Figure 6-10.)

Any examination of controlled ejection (where orientation is retained at ejection) must be made in the light of considering the ejector station of equal importance to any insertion, joining, or gauging station on the machine. This becomes increasingly important with the ever developing use of integrated "work cell" plant layouts.

Controlled Ejection

Since the same type of transfer device can be used to eject out as to transfer in, the accuracy of the ejection motions can be quite accurately controlled and consist of any one of pick-and-place, pick-radially rotate-and-place, pick-rollover-and-place, or pick-dip-and-place motions.

After this movement the part can be dropped to bulk conditions, placed on a conveyor whence it might go either to bulk, to some storage media, or to a subsequent transfer operation for further work. Lastly it may be placed in some type of storage tray, pallet rack, or magazine device before proceeding to subsequent operations. This last choice is usually a costly one and one chosen when there is no other recourse to bulk storage (see Figure 6-11). The use of bulk storage, however, is declining as more and more product is moved to conveyors with buffer storage pallets or packaging equipment.

Controlled ejection is, therefore, a transfer operation and subject to the same operating efficiencies and problems as any other transfer system with the exception that the work-holding fixture or receiver component is often a storage container, perhaps of a broader tolerance and lesser quality and one subject to the wear, tear, damage, and contamination of any shop tool.

Options to X-Y Coordinate Storage Devices

For most of us our first introduction to an X-Y storage device was our discovery of an egg carton in our mother's kitchen. This discovery was often followed by the sound of something fragile breaking and mom's comment, "Now

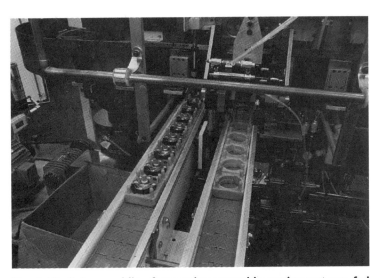

Fig. 6-10. Subassemblies from prior assembly equipment are fed to a final machine using free pallets as carriers. Such free pallets offer a great deal of buffer storage between subassembly and final assembly systems. When laying out such conveyor systems, operator access is a paramount consideration.

Fig. 6-11. Ordnance components must be contained securely once explosive primers are put in place. Aluminum carriers received subassemblies and then go to carrier trays for transfer to subsequent operations.

you've done it!" Most of us shed our childhood behavior to some degree, but fascination with and utilization of X-Y coordinate storage devices for fragile parts is one most hard to discard. Attempts to incorporate such storage systems on automatic assembly machines can lead to a much more vulgar version of "Now you've done it!"

The problem with transferring into (or out of) X-Y coordinate trays is that simple single-plane movement of the transfer device now must become multiplanar or the storage trays must be moved in two planes beneath a fixed transfer point. As each row is filled, index is along a single plane until it is necessary to start another tray or row when movement is required in two planes. A multiplanar transfer device is considerably more complex in design and construction. In high cyclic operation, maintenance and wear are significant. When trays are moved in both coordinates, inertia becomes a design problem when returning to the start position for filling a new tray. *Such stations rarely are able to utilize standard hard-*

ware and become major design and debugging tasks in the development of a machine. The cost of manufacturing such trays can be very expensive and their maintenance a gnawing problem.

There are several ways to get around this problem. One is to make storage trays such that they move along a single axis and are fed to and from a carrier device, which is used to transport these single-axis magazines. One such device used in the ordnance industry to transfer explosive parts is shown in Figure 6-11.

A second means is to produce some form of bandolier (much like belts used to feed machine guns) that can be indexed past the transfer point on a single axis. Figure 6-12 shows such a system. Often the bandolier can be composed of a series of nylon molding thus providing linkage between each pocket because of the elastic nature of nylon.

The circular storage device with pockets machined in an Archimedes spiral described in the section on spring winding is another type of stor-

Fig. 6-12. The use of nylon moldings as carrier chains which may be wound up on reels in an inexpensive storage system that can provide dense storage while stored parts are indexed in a single plane past the transfer point.

age device eliminating the need for multiaxis positioning.

Any design effort that is made to eliminate multicoordinate movement on an assembly machine is beneficial to the operator.

Before leaving this topic, it should be mentioned that the use of plastics as a storage system may be harmful if the accumulation of static electricity will be harmful to the product or to its subsequent operations. Trays and magazines that are subject to dragging may be wound up on reels in an inexpensive storage system that can provide dense storage while stored parts are indexed in a single plane past the transfer point to avoid the sliding motions and static charge buildup that may become significant. Metal carriers more readily shed such charges. Where such charges occur, the use of ionizing devices may prove helpful.

Storage Conveyors

One of the first pitfalls of early attempts at large-scale integration of manufacturing operations is that of providing insufficient float or buffer storage between separate operations. In other words, relatively short conveyors tied each

subsystem together with inadequate provision for buffer storage in the event that any subsystem went down; any subsystem that shut down stopped the whole manufacturing line in a short period.

The most inexpensive way to provide such float (if the product can accept it) is to have each subsystem eject its production to bulk storage containers, which then is subsequently reintroduced to parts feeding, orienting, and transfer devices. Maximum float and minimal storage device costs must be counterbalanced by the very real possibility of product damage and contamination and the cost of reorienting a part that was previously oriented.

The use of storage trays, bandoliers, and magazines provides great product safety in transit and better batch control. The cost of these devices while small per unit of storage can be quite high when one considers the number of storage magazines to be used and the necessity of maintaining these containers in good condition.

Storage conveyors and accumulators are intended to handle the storage of assembled parts which by their nature should be in continuous transit from operation to operation. When well designed, they provide for necessary float in the overall line by storing the output of one subsystem when a subsequent subsystem is down. This ability to store production bulges that cannot be absorbed in the system and having the corollary ability to feed one subsystem when a preceding subsystem is down is essentially done by length (or in the case of rotary accumulators by area) of storage.

In selecting the storage system some hard decisions must be made as to the nature of probable downtime within the system. Will it be a high-incident, low-average downtime problem so typical of high-speed assembly lines? In this case a fairly long conveyor of 10 or 12 ft. (3 to 4 m) may provide a few minutes of float. High-low level controls may be required to prevent machine operation at either end of the conveyor in the event of unacceptable part levels on the conveyor.

If the pattern of subsystem downtime is to be one of less frequent occurrence, but longer downtimes, typical when reels of stock must be changed, or tools or electrodes changed and adjusted, cylindrical accumulators may provide 8 to 10 minutes of surge capacity. Both types mentioned above assume normal operational stoppages as a pattern of operation.

When the percentage of downtime becomes critical to overall production goals or where maintenance, major adjustments, calibration tests, or accidental tool breakage are expected, silo-type elevator storage towers or overhead storage systems can provide significant reservoir capacity. That shown in Figure 6-13 provides 30 minutes storage between systems.

When choosing to incorporate such storage conveyors, the assembly systems customer must be aware of two operational problems. One is to question (and receive a good answer) on how the assembly system attendants can readily clear jams occurring within the storage system, often at elevated heights. "Oh, it won't jam," is not a good answer.

The second operational problem often overlooked is the inability of such a storage system to transfer parts from a preceding to a subsequent subsystem rapidly when the parts reservoir is empty. There is such emphasis on the ability of

Fig. 6-13. Significant float or buffer storage can be achieved through silo-type conveyors. Part motion on this conveyor is achieved by vibratory action.

the reservoir to store from one system or to feed from storage to a subsequent system that one can overlook the problem that occurs when the second system has drawn out all of the available storage and now looks to direct feeding from the now hopefully operational preceding system. Is there provision to bypass an empty storage reservoir or must the second system shut down till adequate storage is built up in the reservoir?

SUMMARY

Ideally, once a part is created, its orientation should be maintained throughout subsequent manufacturing processes. The nature of the products may demand this retention. In the real world of automatic assembly the apparent high cost (vibratory feeders, transfer units, and escapements)

of reorienting the part may appear in a favorable light compared to the capital costs and possible loss of production that may come from attempting to retain orientation throughout its manufacture.

This attitude is changing, however, as quality concerns and the increasing utilization of matched assemblies make it imperative to fully control a developing assembly throughout the entire process. The enormous emphasis on integration has caused enormous development of buffer storage and palletization techniques.

Integration of manufacturing operations must realistically face up to the actual productivity of each element of the integration, provide float adequate to meet the needs indicated by the nature of the probable downtime, and provide some alternative means of continuing production in the event of catastrophic subsystem failure.

Chapter 7
Joining and Fastening

INTRODUCTION

In the chapter on product design, it was emphasized that in many cases optimal assembly design should eliminate any requirement for specific joining or fastening operations. Snap fits, twist locks and interference fits often can be handled as part of part transfer functions. There then is no need for specific joining stations.

In any gradual retreat from optimum conditions, we must look beyond assemblies joined with interference fits to those assemblies that are staked together by deformation of one or more component parts. The next stage of sophistication involves joining or bonding of two or more components by the use of heat, pressure, or relative movement. Welding, ultrasonic welding, and spin welding are typical of this type of joining. Lastly, joining may be done by introducing a bonding agent such as solder or adhesives or catalysts, or the use of mechanical fasteners such as screws, rivets, and the like.

While almost all conventional and emerging joining technologies can be applied to assembly machinery, it must be recognized that in many types of joining, line balancing requirements can be a severe problem. Joining-station duty life may be more severe on an automatic machine than on

a battery of hand-fed single-station joining machines.

The beneficial aspects of joining on assembly machines can include greater uniformity and consistency. Isolation of the feeding function and fastening function of mechanical fasteners, possible only in multistation assembly systems, particularly if product design has recognized this capability, will greatly reduce scrap generated in the mechanical fastening process.

Lastly, no consideration of joining can ignore the special problems of wire termination. Major development in assembly of electrical devices will come from entirely new approaches to wire termination and mechanization. Unfortunately, many of these terminations may prove inadequate in the product's work environment. Welding or soldering may have to be used, particularly when the product will operate under severe vibration.

The decision to mechanize assembly will almost certainly force a review of joining practices, particularly where present design includes joining methods having severe line imbalance with the other assembly related functions. To illustrate this, some years ago a major producer of fingertip detergent sprayers decided to mechanize the assembly of this product. Under the then existing design there were three joining operations. The plastic components were bonded through the use of solvents. This technique required a minimum of 24-hour cure time between each bonding operation. Total assembly time for a simple pump often ran 1 week or more. In-process inventory often exceeded 500,000 assemblies. The decision to mechanize led to an investigation of ultrasonic welding. A change in materials allowed ultrasonic welding. Cure times were eliminated. Dermatitis problems and fire hazards were also eliminated. In-process inventory was reduced to 35 parts. Dramatic inventory reduction alone paid for the assembly system. Greater uniformity of joining eliminated field problems of inserting the pump in high-speed bottling lines. There are many such dramatic examples.

INTERFERENCE FITS

The use of wooden pegs or nails to join parts is the earliest joining technique. One part is driven into another. Providing the cavity is identical to, or smaller than, the part being inserted, friction between the two parts will inhibit disassembly. The implicit assumption, of course, is that one or both parts have some degree of malleability or elasticity allowing the more malleable part to conform to the more rigid part. This conformation does not always exist between different materials to the desired degree. The frictional forces may not be sufficient to prevent disassembly under normal loads. There is always the danger that the recipient part may not have sufficient malleability to receive the inserted part without fracturing. In simple terms, press three thumbtacks into a piece of pine wood, into a piece of soft dry wall, and into a hard chip board. In the first case, adhesion of the thumbtack is sufficient for all normal uses. In the second case, the adherence is so slight that the least load will cause the thumbtack to fall out. In the last case, an attempt to force the tack into the board will either destroy the tack or fracture the chip board.

The obvious advantages of assembly by use of interference fits are reduced component costs, lower capital equipment costs, and an absence of line balancing problems. The critical design areas are the suitability of the component materials, particularly in the aspects of column strength, malleability, and tensile strength, and the component fabricator's ability to hold part tolerances within limits that will not cause destruction of parts or failure to provide sufficient friction.

The use of barbs on fish hooks, spears, and arrows was an early recognition of this problem. Our present use of grooves in dry wall nails is a modern attempt to increase friction between two parts. If that is not sufficient, we may consider various means of increasing friction by post-insertion deformation of one or more parts through the use of pressure.

JOINING BY PRESSURE

Controlled pressure is the basis of a number of processes used to join assembly components. Interference or press fitting, swaging, staking, and crimping are representative examples of these operations. When such joining operations are incorporated in complex assembly machinery, attention must be given not only to the method used but also to its effect on machine costs and efficiency.

In this examination of various methods used on assembly machines to join components by pressure, particular emphasis is placed on the relationship of the joining operation to the total machine operation.

From this viewpoint, the type of unit used for upsetting, staking, or forced insertion must be considered on the basis of its costs, its operational dependability, and its effect on other machine operations. The assembly machine builder has three special concerns:

1. The isolation of impact or vibration of the joining operation from other transfer and insertion stations.
2. The consistency or uniformity of the joining operation.
3. The safety of the machine and operating personnel.

Isolation of Pressures

How should pressure be applied in the assembly process? The answer to this question is closely associated with the basic type of assembly machine chassis chosen. The basic chassis selected must provide access for press anvils that can withstand high ram or impact pressures. To the fullest degree possible, the station design should be such that ram pressures are not transmitted to the machine frame. (Figure 7-1.) Transmission of ram pressures through the machine frame increases wear on the machine chassis and tends to destroy the index accuracy of the basic mechanism.

Vibration caused by impact deformation also may be important if it would tend to displace the assembled parts that have not passed through the joining operation, or to destroy the efficiency of inserting stations on the machine. When there is a possibility of such displacement, pressure applied by a squeezing or compression action is preferable to that created by an impact.

Consistent Deformation

One of the most common problems in joining components by pressure is the lack of uniformity in upsetting or swaging operations. This condi-

Fig. 7-1. A "C" frame on this hydraulic press is designed to eliminate stresses on the basic machine system.

tion is generally caused by variation in the amount of pressure applied or variation in the dimensions of the components to be assembled. In the first instance, variation in the pressure applied is often a result of improper installation or maintenance of the press equipment, pressure fluctuations in the pneumatic supply line, or overheating of the operating fluid in hydraulic-operated presses.

The most prevalent joining problems in mechanized assembly through component deformation are those caused by variation in the length of the parts to be assembled. Here, the difficulty lies in providing adequate deformation or insertion without crushing the components. If part quality is held to levels consistent with economical manufacturing practices, the effect of acceptable length variations can be largely overcome by controlling the joining station's operation on the basis of a predetermined pressure rather than a mechanical stroke limitation. In this way, the tolerance buildup in individual sets of components controls the action of the machine. Hydraulic presses are particularly well suited to control of stroke by pressure rather than by a fixed dimensional limitation. However, it is also possible to incorporate pressure control in mechanically actuated rams.

Safety of Personnel and Machines

Occupational safety laws are very specific in spelling out the types of guarding needed around hazardous work stations. Also, conscientious plant safety departments should be counted on to provide for personnel protection through instruction on the use of proper guarding equipment. Nevertheless, the assembly machine designer must take all steps to prevent accidental operation of rams during those times when guards are removed, such as machine setup periods or when jams are being cleared. From a control standpoint the question of operator safety is closely allied to the problems of ensuring that press or riveting operations occur only during the proper portion of the machine cycle and that inspection devices are included to determine the position of all components prior to the joining operation. The hydraulic installation shown in Figure 7-2 is an example of the provisions that

Fig. 7-2. In this hydraulic press station, single-cycle controls are key switches operable only by trained maintenance workers while interlock switches provide for actuation only when all components are properly placed.

must be made against accidental actuation of press rams.

Effect on Productivity

Impact deformation or staking has little effect on machine cyclic rates. Squeeze-type deformation is a somewhat slower process.

Spinning or vibratory riveting requires specific time periods for material flow. This time period must be determined accurately and compared to other machine limitations. When the length of time required for proper material flow seriously limits the inherent productivity of other work stations on the machine, step-forming operations can be performed much in the same way that step cutting is done on multiple-spindle drilling equipment. That is, the spinning or upsetting operations may be distributed across two or more work

stations to bring the cycle time requirements of each of these operations into the same range as those for the other work stations on the machine. However, such step-forming techniques cannot be applied if the material is work hardening to such an extent that all of the forming must be accomplished at one time. Certain materials may work harden during the first forming, making a second forming operation much more difficult.

Cyclic capability of control system relays or hydraulic valves may also impose upper limits on machine rates. The assembly machine designer must also be aware of those limitations imposed by the material or by inherent characteristics of the equipment when quoting production rates. For instance, time required to recharge capacitors or power supplies may limit cyclic rates for certain types of lasers, welders or magnetic forming units.

Methods of Exerting Pressure

A wide range of methods can be used for inserting parts or deforming materials. Some of these techniques are more properly suited to metal deformation. Others are best for inserting operations. However, the majority of the methods are equally suitable for both types of work.

These methods all have their problems. In some instances, noise of operation may be a vital consideration. This subject is one of continuous concern in many large manufacturing firms. There are also the special problems that stem from mechanization. Much skill goes into good manual joining operations. This is particularly true in riveting. When manual operations are replaced by machines, the methods of control must be carefully selected so that the joining stations are capable of compensating for acceptable manufacturing variations in the components. In the following discussion, the most commonly used methods of joining components by pressure are examined in the light of these criteria.

The Machine Slide. Assembly machinery with vertically moving tooling platforms can use the downward movement of the upper tooling platform for exerting light pressures. This is an inexpensive method of using a preexisting motion.

Similarly, a toggle-type linkage may be connected to one of the tooling camshafts of an integral assembly machine to exert pressure. In both of these arrangements, the pressures required must not exceed machine limits. Otherwise, fatigue failure, galling, or other damage to the basic machine mechanism may result.

Such devices are mechanically simple, easy to synchronize, and quiet in operation. They cause little vibration in the adjacent stations. But, if designed as shown in Figure 7-3, they have three inherent limitations.

The first limitation is a relatively fixed mechanical stroke. The stroke setting is always a compromise between the extremes of inadequate forming and crushing of the assembly. However, for the unit shown, this condition was not a severe problem. Only one component is being deformed, so only the small variation in gauge of metal is involved. Also, in this particular assembly, some movement was desired between the two components. In more complex assemblies, variations in components can cause some of the problems mentioned previously.

The second limitation is one of machine safety. If the movement of a unit is mechanically positive, presence of some foreign object under the press ram can cause serious damage to the equipment. A common occurrence is the failure to strip away an assembly from the punch or ram of the press. When this happens, the assembly is carried up by the ram and, following index, is carried down on top of the next subsequent assembly.

The third limitation is the possibility of improper alignment or positioning of the individual components relative to one another prior to applying pressure.

To prevent damage to the equipment or components, three steps can be taken. The first step is the inclusion of an inspection station, prior to the press station, to check for presence and proper relative position of components. The second step is an electrical inspection unit on the press ram to ensure stripping of the punch. A small detector pin, which may or may not actually aid in the stripping, can be inspected during the index portion of the machine cycle. If its position indicates incomplete stripping, the machine can be stopped automatically. The third step is to change the design of the punch from a fixed mechanical stroke to a pressure-limited stroke. This can be accomplished by placing a compensating device between the machine slide or camshaft and the ram itself. Generally, this device is some form of spring; however, one-way pressure-regulated air cylinders can also be used. When a spring is used, it must be preloaded at the maximum ram stroke to the desired pressure. If longer acceptable components cause further deformation of the spring, there is some increase in pressure but no damage to the machine. (Figure 7-4.)

An important machine design consideration is the question of whether or not sufficient compensation can be built into the unit to prevent machine damage purely by overtravel. This may require, in the station design, compensation for the full length of the component being inserted or deformed, over and above the normal operat-

Fig. 7-3. Mechanical toggle presses are driven from the main camshaft.

ing position of the punch. When Belleville-type washers are acting as pressure-limiting springs, such overtravel is often hard to obtain. In this event, the wisest solution may be to electrically inspect for excessive compensation of the ram, stopping the machine upon detection of such a condition.

Toggle Mechanisms. The mechanical press is basically a toggle mechanism whose maximum force is exerted at the moment the linkage pivots are aligned. Assembly of separate components presents different problems for a mechanical press than does the piercing, blanking, or drawing of material. Here again, it is the question of whether or not a fixed mechanical stroke is able to handle the permissible manufacturing tolerances of the individual components and still produce a uniform deformation or insertion. Since the power input to a mechanical press is continuous and rotary, the press must always follow through to its maximum preset stroke.

Fig. 7-4. Between the punch and punch housing a group of Schnorr springs act as a pressure-limiting device.

Although many small assemblies are produced on dial-fed mechanical presses, synchronizing such presses to complex assembly machinery is difficult. This is particularly true when the assembly machinery is running relatively slowly and a direct coupling arrangement of the press to the machine's camshafts would destroy the inertial advantage that such toggle presses have while running at higher speeds. If there can be no direct coupling, some form of press clutch must be used.

Air-operated toggle presses use the same type of mechanical buildup through a crank linkage. However, the power input is a reciprocating motion and the stroke can be limited by the pressure input to the operating cylinders. By virtue of the nature of a mechanical toggle linkage, a geometric increase in pressure is exerted as the toggle linkage swings toward center. Wide variations in tolerance buildup of the individual components of the assembly will, in turn, affect the pressure capability of the toggle unit itself.

Fluid Power Presses. A gentle impacting squeeze type of pressure can be derived directly from the ram pressure of large air or hydraulic cylinders. Since air pressure in shops is generally below 100 psi and rarely over 120 psi, high press ram pressures can be obtained with air cylinders only by increasing the cylinder diameter. A hydraulic press similar to those illustrated in Figures 7-1 and 7-2 can provide the high pressures that may be required. Hydraulic presses are often blamed for erratic operation when actually the cause is an overheated oil supply. When coupled to automatic machinery, such presses should preferably have proper valves adjusted so that the forward stroke is stopped upon reaching a specified pressure, rather than a specified linear travel. Also, the valve arrangement should provide momentary pause before return of the ram. This allows the punch to remain in position long enough to allow material flow in the assembly to stabilize. Even a few milliseconds pause is helpful. In addition, the valving should allow pressure to drop quickly, or pump discharge to be dumped back into the reservoir, whenever high pressures are not required. Such relief valving will help to keep the oil temperature down. If the cyclic work rate is heavy enough, the oil reservoir may have to be cooled by heat exchangers.

In all cases where fluid-power cylinders are used for obtaining pressure on mechanically integral machines, the ram should be electrically interlocked so that the machine does not attempt to index while the rams are forward.

Impact Hammers. In the fluid-power presses previously described, pressure is derived completely from the force of a compressed fluid against the cylinder piston. Air impact devices release this pressure to the cylinder all at once, causing the piston to shoot forward like a projectile. The ram, accelerated by this action, shoots forward until it strikes the part. The work is done by the momentum of the ram and not through the lingering pressure that is exerted against the piston once forward movement has stopped. Large air impact units by their action are able to derive a much higher rate of deformation than would be possible through pressure exerted through the cylinder alone.

Electrical Impact Staking. The action of an air impact hammer, which uses the energy of an accelerating mass, also can be obtained electrically. Figure 7-5 shows an electrically powered impact staking tool, swaging hubs of hypodermic needles. In this unit, a free floating core is held at one end of an electrical coil. Energizing this coil suddenly pulls the core forward while de-energizing the coil permits spring return of the ram. The force of impact is controlled by the amount of current passed through the windings of the coil. Thus, a simple rheostat adjustment can be used to control the amount of pressure exerted. These units are capable of high-speed operation. Approximately 0.025 second is required for a single controlled blow. With proper controls the unit can be programmed to strike repeated blows at one index of the machine.

Spin Riveting. Several methods may be used to provide a gradual deformation of material under pressure. The cycle time associated with these methods requires special consideration. One method that produces excellent finishes and that is extremely quiet is spin riveting. Figure 7-6 shows various spinning tools used for spin riveting. These units are readily adaptable to multiple-station automatic assembly equipment. Spin riveting uses two split rolls, shaped to the desired contour of the part to be spun. When appearance is critical, such units will produce an extremely smooth finish.

Gradual upsetting also can be obtained by the repeated impact of a vibratory-type riveter. The repeated high-frequency blows are able to displace proportionately large amounts of material in relation to the low air pressures required for operation. The main problem in incorporating such equipment into an automatic assembly machine is the difficulty in ensuring uniform or complete upsetting of the material within the machine dwell time.

The effects of both types of deformation and the advantages of spin and vibratory riveting are combined in a single unit, Figure 7-7. In these units, a peen vibrating along its axis is turned in a rotary motion at the same time. The rotating

Fig. 7-5. An electrically operated impact hammer utilizing a solenoid-driven motion.

Fig. 7-6. Various spinning tools for spin riveting operations. From left to right: wobble tool, crimping tool, split roll spinning tool, sealing tool for tubes, curling tool.

motion can be obtained with an air motor or an electrical motor, Figure 7-8.

Vibratory spin riveting has a characteristic surface finish, somewhat textured. While functionally an excellent finish, this appearance may not always be acceptable in consumer items.

Orbital Riveting. Much of the work formerly done by various types of impact hammers, spinning-type riveters and impact or squeeze-type riveters is now being done by riveters referred to as orbital riveters. Although this type of riveting has been available for many years in the United States, its present broad-scale use must be attributed in great part to successful marketing of this technique by several European manufacturers.

Impact deformation in riveting often causes splitting of material and is noisy. Vibrating riveters share this noise problem. Hydraulic presses used for riveting and forming are also noisy and require energy not only for operation, but energy for cooling. Air-operated presses depend on costly compressed air and may be erratic owing to air-pressure fluctuation.

Spinning-type deformation is the quietest type of deformation and has the lowest energy cost. The material deformation that occurs is done gradually and can offer improved metallurgical characteristics. The use of spinning-type deformation is an old art. Typical examples are rivets used to secure wooden and plastic handles to cutlery. Cosmetic appearance is critical. In the earliest forms rolls shaped to the desired rivet are split at the centerline of rotation and rotate in opposite directions on a common axle pin. This type of rivet deformation is extremely useful when flush riveting and appearance are critical. The assumption here is that product design and fixturing will allow the spinning tools to rotate without interference.

Newer spin riveters eliminate the split rolls for small peening tools that rotate under axial load to form rivets, lugs, or casings to desired shape. In some cases the tool simply rotates, and in other cases it moves in an orbital pattern to produce fi-

Fig. 7-7. Vibratory spin riveting combines rotary motion and impact hammering.

nal form. Each manufacturer makes claims for their specific pattern of movement. In general, they all work exceedingly well and are ideally suited for inclusion in automatic assembly equipment. (Figure 7-9.)

The key elements to be considered in specifying or approving any system using pressure for joining, even if that axial pressure is greatly reduced by rotation of the forming tool, should include:

• Energy (and cooling) costs
• Noise levels
• Provision in the tool design for isolation of pressure and vibration from the indexing mechanism, machine chassis, and if possible the fixture carrier
• Electrical (or optical) detection to ensure that deformed parts are not picked up in the ram, or forming tool
• Design features that ensure the forming tool will not damage the workholding nest if for

some reason a work-holding fixture indexes into the joining station without component parts being present.

In almost every transfer station used to feed or insert parts, safety compliance can be included in the design to prevent damage to machine and fixture. In pressure-type joining, however, significant pressures are exerted in normal operation with the expectation that they will deform product components. If for some reason these parts are missing or improperly positioned relative to one another, the design of the machine and primarily its control system must prevent accidental damage.

It is important to emphasize once more the beneficial aspects of separating the placement and deformation of fasteners wherever practical. The sequence of feed-inspect-deform is more costly than insertion and deformation in a single station, but the extra capital costs should quickly be recovered in reduced downtime and maintenance costs. More importantly, scrap reduction, particularly in products joined by multiple rivets, may be a significant factor in machine payback. This separation will require that product design provides for maintaining the fasteners in proper relative position during index to the inspection and joining stations. This same consideration applies to threaded fasteners examined later in this chapter.

In our discussions to this point of joining by pressure, no distinction is made between those assemblies in which one component is crimped or deformed to hold the assembly together and those in which rivets are introduced as additional components in the assembly.

Functionally there is little difference, but, in terms of product cost and machine efficiency, elimination of fasteners is usually preferable. Assembly by pressure has one other drawback. It usually precludes product disassembly for full service, adjustment, and repair. Where the ability to disassemble and reassemble is vital, threaded fasteners must be used.

Before leaving the area of riveted construction, it might be mentioned that the parts feeding techniques for screws and rivets are for the most part identical.

Fig. 7-8. An electrical motor produces motion of the peen.

THREADED FASTENERS

Threaded fasteners are fed and inserted automatically by the billions each year. This section examines automatic assembly techniques for threaded fasteners.

Headed Screws

Any screw with a flanged head section larger than the thread diameter presents different problems of feeding and insertion than do set screws or studs. The methods described here apply equally well, with some slight modifications, to machine screws and self-tapping screws; to socket head, Phillips head, and slotted screws; to fillister head, button head, and flat head screws; and to hex head bolts.

The three most commonly used units for feeding headed screws are barrel, blade, and vibratory feeders. In general, screws are easily oriented and fed, particularly when the ratio of body length to head diameter is generous. As this ratio approaches unity or even less, it becomes increasingly difficult to select the screws with the head up, especially in vibratory feeders. Turnover devices, pendulums, and strategically placed air blasts are used to orient the screws to an acceptable position.

Once the screw is oriented, one of two methods is used to feed the fastener to the screwdriver.

Fig. 7-9. An orbital riveter with micrometer adjustment.

When the screw-inserting spindle is in a fixed position and access to the spindle is relatively good, the screws are most normally fed down inclined rails. When the inserting spindle is to be moved or access is poor, the screws are escaped and blown, one at a time, to the inserting spindle, through a flexible feed tube.

Several methods may be used to transfer the screws and to hold them in position during the inserting operation. One commonly used method, Figure 7-10, is suitable for both fixed-position and flexibly mounted spindles. Here, the screws are escaped, or blown, one at a time into spring detented fingers. These fingers have an opening large enough to allow the body of the screw to pass through but small enough to retain the head. As the spindle is activated, the screwdriver pushes the head of the screw through these fingers, which are cammed open at the appropriate time to allow the screw head to pass through.

Another method, Figure 7-11, is suitable only for a fixed-spindle position. Here emphasis is placed on driving the screw head while the screw is held by its thread.

When any screw inserting system is evaluated, the actual method of screw manufacture should be considered. Generally, screws are upset, or headed, from a wire blank, then roll-threaded and finally slotted. This is generally done in three separate operations, and some runout between the thread and the screw head is normal. Also, there is generally some discrepancy between the centerlines of the thread and head diameters, and the location of the driving slot. These discrepancies, if significant, can seriously affect the inserting efficiency of automatic screw-inserting machinery. *The time lost from production and spent in salvage operations more than offsets any price advantage gained in procuring substandard screws.* Unfortunately, *screws are often purchased as a generic commodity with little or no regard to assembly machine efficiency or product quality.*

Screw-feeding systems holding the screw by the threaded portion will generally be more expen-

Fig. 7-10. A screwdriving unit which retains the screw by its head during initial screw engagement.

sive, but have greater operating reliability than those systems holding the screw by the head or flange.

Often, subsequent assembly operations dictate the need for controlling depth of screw insertion. This is generally coupled with other specific requirements. On electrical binder screws, for instance, it is customary to insert the screw to full depth, to upset or spin the tip of the screw to prevent accidental withdrawal or loss, and then to back the screw out, so that the electrician may easily place the wire around the screws. For this operation, the back out spindle is driven through inertia and the screw stops when the staked or spun portion hits the other threaded component.

In other applications, such as brake-adjusting screws, headlight-adjusting screws, and valve tappets, it is not a question of backing out a screw, but rather one of inserting the screw to a determined number of turns so that adjustment will require a minimum amount of time and attention at final assembly or product calibration. Here, a reversing or tapping spindle having a unidirectional or free running clutch between the spindle and screwdriver is used to insert the screw, Figure 7-12. The number of turns, or

length of stroke, on such spindles is easily set, and the clutch prevents screw backup when the spindle is withdrawn. This type of installation is quite accurate within the limitations imposed by the manufacturing tolerances of the components.

The relationship of the screw threads to the driving slot, the variations in lengths or heights of the components, and the angle that the screwdriver must turn through before engaging the driving slot or cavity will determine how accurately the depth of insertion can be controlled.

When depth control is important, the engagement of thread should be sufficiently tight and the inserting spindle speed should be slow enough so that inertia does not carry the screw from the screwdriver and past the predetermined limits. If necessary, a small dab of grease on the threaded hole may reduce this excessive insertion of the screw.

Headless Set Screws

The actual insertion of set screws is no different than for any other screw; in fact, problems of runout are lessened. But the orientation of set screws prior to insertion is more demanding than for headed screws.

Fig. 7-11. Screws are held by the body, providing better concentricity during initial screw engagement.

Specific equipment for feeding and orienting set screws has been developed by some manufacturers for their own products. Several of the methods have been patented. However, most set screw feed systems use one of four basic methods. They are:

1. Selector wheels
2. Counterrotating rolls
3. Keyed tracks
4. Floating ring selectors

The Selector Wheel. The selector wheel has two basic variations. In one type, shown in Figure 7-13, the periphery of a small intermittently indexing wheel has a series of holes that are the same size as the screw to be selected. A pilot that is slightly smaller than the driving slot or socket is located at the bottom of each hole. Screws are fed randomly end-to-end to this wheel. If the screw is oriented so that the driving slot or socket comes first, the screw completely enters the cavity. As the wheel indexes, the screw is carried around 270° until it is free to fall into a transfer tube with the driving slot or socket upward. If the screw enters the wheel point end first, the pilot prevents complete entry. Strategically placed air blasts would then blow the screw out of the wheel and back into the feeder. With this system, only relatively short screws can be selected.

In the second variation the peripheral holes in the selector wheel do not have pilots. Screws fed end-to-end enter the holes in the wheel with either end out. As the wheel indexes, the screw is carried under an inspection probe. If the probe senses that the driving slot or socket is uppermost, a gate allows the screw to fall through to the transfer tube. If the probe showed the other end uppermost, the screw would be carried around to be returned to the bowl for refeeding.

Fig. 7-12. A unidirectional sprag clutch mounted on a controlled reversing spindle is used to drive screws to a known depth.

Here, screw length is not important in the selection process.

Counterrotating Rolls. A second method used for orienting set screws uses two counterrotating hardened rolls at the discharge end of the vibratory feeder. These rolls have an orienting groove that allows the screw to drop through to the discharge tube. Set screws tend to have one end heavier than the other. Screws, coming to the rolls with the wrong end foremost, pass over the orienting groove until the heavy end can drop in first. In most instances a slight taper or substantial chamfer on the outer diameter of the screw is needed to select the screws properly.

Keyed Tracks. When selecting set screws having a driving slot, whose length is almost the same as its width, a keyed track can be used effectively and is simple in construction. The feeder track has a small key, or strip, just slightly smaller than the width and depth of the driving slot in the screw. As the screws are fed, some will drop over the key, and a gate fastened over the feeder track

will permit these screws to pass while screws in any other position would be returned to the bowl of the feeder. Once the screws are passed through this gate, they are retained in a track that is twisted, turning the proper side up for screw-driving operations.

Floating Ring Selectors. This method for selecting headless slotted set screws is based on the floating ring vibratory feeder principle. Here, a free floating ring turns in a machined slot on the vibratory feeder bowl. Screws feeding up the spiral track of the feeders move out across the top of the ring and drop at random into a series of holes machined in the ring. The vibration of the bowl causes the ring to turn freely in the direction of the bowl rotation. As the ring turns, it carries the screw in the holes over a cavity machined in the feeder bowl. If the screws are properly positioned (driving slot or socket upward), a blade catches the last uppermost thread in the set screw and retains it. The screw is then carried by the ring to a discharge tube where the screw falls to the inserting fingers. If the screw passes over the cavity with the point upward, it is in a position that prevents the blade from catching the last thread, and the screws fall into the return cavity for another trip through the feeder mechanism.

The floating ring selector can be used for either slotted or socket-type set screws, providing the socket set screw has a substantial chamfer or dog point. It requires no drive or control mechanism other than the vibration of the feeder. In addition, the system is not limited to any specific length of screw or diameter to width ratio.

Stud Driving

The main problem with completely automatic stud driving is the difficulty in selecting the proper end of the stud. Generally, studs have one coarse and one fine-threaded end. Without a flange or significant weight difference between the two ends, it would be extremely expensive to select these studs for automatic driving. When studs are used, they are generally applied in a large number, and one usual assembly technique is to start the studs by hand and then use a gang, or multiple, driver to set all the studs to depth at one time.

Fig. 7-13. A pinwheel-type screw selector. Screws coming up the vibratory feeder track attempt to enter cavities in the intermittent selector wheel. Those properly oriented enter, while others are returned to the bowl.

These stud drivers are automatic in operation. Their construction usually consists of a center plunger that is depressed by the stud when it is properly inside the driver.

Depressing this center plunger closes the driving jaws on the threaded portion of the stud. These jaws are released when the stud is driven to a certain depth, or when a certain torque requirement is met. If the number of studs to be driven is small enough so that feeding efficiency will not result in undue downtime, completely automatic driving can be a practical method.

Fully automatic multiple stud drivers for items such as automotive blocks are highly standardized.

Nut Running

In general, nuts are more difficult to feed and insert automatically than screws. The problem is that their shape usually makes it difficult to hold and drive them at the same time. If not started properly, nuts tend to cross-thread, damaging both the male and female threads, and a defective assembly is the result. Whenever practical, a very strong attempt should be made in the design and layout of assembly machines to feed the nut first and then drive the threaded part or screw into the nut. This is not always possible, of course.

The best method for successful and efficient nut driving uses two spindles. In the first, or inserting station, the nut is held squarely in transfer fingers and driven by friction just enough to ensure that the threads engage squarely. The part is then released from the transfer fingers and indexed to the next station. Here, a more solid spindle engages the flats or driver slots on the nut and drives it to the proper position and torque. Figure 7-14 shows a machine that assembles electrical conduit connectors and drives both hex nuts and electrical lock nuts.

The final chapter in this book discusses future assembly trends. There is increasing criticism by environmentalists, especially in Germany, of

Fig. 7-14. An electrical-conduit-connector assembly machine assembles hex nuts and electrical lock nuts.

products that cannot be repaired or "remanufactured." It is not impossible in the near future to see greater use of threaded fasteners to permit remanufacture or repair of products. This will have significant assembly consequences.

JOINING BY ADHESIVES, CATALYSTS AND SOLDER

Adhesive Bonding

Say "glue" to some adhesive specialists and the results may be anything from raised eyebrows to outright belligerence. Say "adhesives" to an assembly machine designer and the results will usually be an audible groan. One popular song in World War II was the lament of a young girl about available men on the homefront. They were either too young or too old, too short or too long. For most assembly system builders the use of adhesives in assembly is much the same. They are too viscous or too watery. They set up too quickly or too late; and to add to the confusion, technical assistance is often in short supply. As the complexity of various epoxies and adhesives increase, far too many manufacturers leave the application of their product in limbo, as if all assembly work is done by the elves in Santa's workshop. Some few companies involved in chemical bonding such as the Loctite Company deserve special thanks for the complementary development of modern day dispensers like that shown in Figure 7-15 adaptable to mechanized assembly systems, and the technical support given to product and process engineers.

The assembly machine designer has many problems with adhesives. One major problem is a continuing concern with accidental contamination of fixtures, transfer tracks, and placement units. *Any design review must consider if and how the problems of fixture contamination have been addressed.*

Another major concern about adhesives is the common necessity of clamping parts after joining to allow the adhesive to set up. The use of clamping-type fixtures to hold the part compressed until the bonding agent begins to take hold is usually to be avoided, particularly in synchronous machine systems. Fixture costs sky-

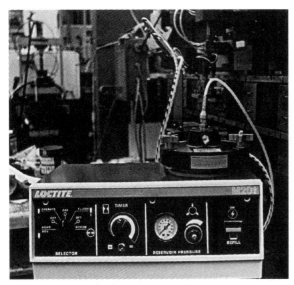

Fig. 7-15. A modern adhesive dispenser provides self-purging capabilities.

rocket. Inertial problems increase. Production rates are inhibited.

Whenever adhesive or epoxy specifications point to long cure times, be certain to distinguish between total cure times and the time required to attain sufficient product integrity to permit unclamping and ejection. In some cases, the total cure time may take hours but sufficient physical integrity for ejection is available in minutes or less.

Can sufficient curing to permit ejection be assisted by heat? Air heaters, infrared lamps, ultraviolet lamps and even calrod units may be able to accelerate the bonding action. Chemical accelerants may be coated on matching parts.

Often overlooked is the ability to design into the product features that permit the assembly to be self-clamping during the curing process. If, for instance, adhesives are felt necessary to ensure the seal of a cover to a housing, a cover to tube, or a diaphragm to a chamber, the need for clamping-type fixtures may be eliminated through the incorporation of snap-in features in product design sufficient to hold the parts physically together without the use of external clamps. These snap-in features will not ensure a hermetic or liquid-proof seal, but will hold the components together while the adhesive or bonding agent is cured.

Lastly, engineers involved in adhesive joining

should be aware that component parts may be coated with an accelerant and then dried to permit automatic feeding. Once this accelerant coating on one part touches another part with adhesive, material curing times are shortened. Extensive investigation into adhesive properties may result in greatly reduced capital equipment costs.

The question of pot life of two-part epoxies or adhesives is of great concern to the builder. Too often, concentration on how the material is dispensed tends to ignore what must happen when the machines stop. Are automatic pressure purging features required in the dispensing system? If this is not practical, can the machine utilize throw away dispensers and nozzles? How do you handle a 10 second stoppage? A 10 minute stoppage? A lunch break? A weekend?

In the past, one of the great challenges of adhesive dispensing was the question of verification that the material was dispensed and placed in the proper area in sufficient quantities. This was particularly difficult when the material was translucent or highly viscous. Some suggested the use of fluorescent dies that could be detected by infrared reflection.

A more practical approach was the use of redundant stations with the emphasis on process reliability rather than process verification. Other approaches included dispensing material through electric eye systems, so that the material passing from the dispenser would trigger photoelectric circuits. Today, rapid advances in optical sensing equipment can detect the presence of adhesives and epoxies because of the change produced in the surface texture of components. This emerging technology will permit better verification of the presence of lubricants, greases, adhesives, and other fluids or semifluids.

Another significant consideration in the use of adhesives involves various occupational safety aspects. Many bonding agents are chemically active. Dermatitis is a major concern. That, coupled with other health hazards, may force the builder to insist on hold-harmless agreements from the customer to reduce the builder's product liability exposure.

Another critical item from a safety standpoint is to determine whether pressure vessels will be required to contain the adhesive epoxy mix. If so, every attempt should be made to physically interlock vessel lids or covers so that it cannot be removed without purging the vessel of pressure first. Having witnessed such an accident during an acceptance run it might be said that while there were certain semi-comical aspects to seeing a major machine glued solid and immovable, several weeks were lost in restoring the machine to operation. It was a miracle that three people were not blinded.

Solvents and Catalysts

We have mentioned the use of accelerants to achieve faster bonding. Some plastic materials are susceptible to solvent bonding. The solvent itself is not the bonding agent, but rather alters or liquefies for a period of time the surfaces of two component parts so that a molecular joining occurs before resolidification.

The use of solvents presents few problems to the machine builder, but severe problems for the system operators. Simple dispensing systems are aided by the capillary like nature of solvent distribution. From an operational standpoint, solvents are usually toxic and flammable and cause severe allergic reactions in many people. From a product standpoint, solvent bonded assemblies often exhibit resistance to tensile forces equal to component strength only to fail quickly when subjected to impact or physical shock.

The advent of ultrasonic welding has replaced the use of adhesives and epoxies in many small parts assemblies. The physical characteristics of parts or materials suitable for ultrasonic welding, however, may not be suitable for product function. Some use of adhesives will always be with us.

Soldering and Brazing

One of the most widely used bonding materials is solder. Alloys used for solders exhibit a wide variety of properties and are capable of liquefaction at almost any desired temperature. In certain applications, this ability to change from solid to liquid is quite functional. Thermal fuses and metallic fire sprinklers depend on this characteristic for their ability to ensure that thermal safety boundaries are not exceeded.

When solder is used to bond products being automatically assembled, there are several considerations for the assembly machine designer to evaluate. The two basic problems in soldering or brazing efficiently are to ensure clean surfaces on the parts to be joined for proper adherence of the solder and, secondly, to ensure efficient transfer of heat to the surfaces to be joined.

In brazing, a commonly used assembly technology is to place solder preforms between the parts to be joined and then place the assembly in a brazing oven where temperatures are closely maintained while a controlled atmosphere reduces or eliminates oxidation on the surfaces to be joined.

From an automatic assembly standpoint, some consideration of brazing or solder preform usage is helpful. In the brazing oven (which may be a batch-type or continuous chain oven) the preform melts and will, in a capillary fashion, move to those areas where component parts are closely mated together. The oven temperature must be closely controlled, and one important control factor is to keep the level of product within the furnace constant. In batch-type ovens, this is fairly straightforward. In continuous-chain-type ovens directly coupled to assembly machines by feeding conveyors, there is a vital consideration. The net production of the assembly machine must be in excess of the furnace capacity to ensure that the load in the brazing furnace remains constant. It will be necessary to provide controls sensing the level on the brazing oven conveyors, which in turn run the assembly machine on demand to maintain proper part levels on the conveyor and in the furnace.

Care should be taken when directly coupling assembly machines to conveyor-fed brazing ovens that sufficient "float" or buffer storage be provided between the two units. Some form of buffer storage or accumulator should be used to accommodate the steady consumption of parts in the brazing oven to the more variable output of the assembly machine. (Figure 7-16.)

Preforms of brazing alloys or solders may be stampings of thin sheet stock (often made on the assembly machine) or may be rings formed from wire-shaped alloy. These split brazing rings may present real problems in automatic feeding and

Fig. 7-16. A horizontal vibratory rail accumulates the output of three dial-type assembly machines to keep the level of parts in the brazing oven constant.

transfer. Most brazing alloy wire material is relatively soft, and when rings do not have their wire ends tightly butted to one another and the gap at the split in the ring exceeds the wire diameter, the rings will interlock like a daisy chain causing jams in part feeders. It is important that the wire ends be butted together as tightly as possible. This condition can often be achieved more readily if the wire rings are formed on machines using straight slide forming such as fourslide machines, rather than on spring-forming machines, which tend to form the ring with a slight helix. This helix increases the chance that gaps which allow interlocking will occur more frequently.

The fragility of solder preforms and the difficulties of feeding such preforms has led to the increasing use of solder or brazing pastes, which may include fluxes. These pastes are easily metered to the proper areas of the assembly.

Other than the concern over fixture contamination from feeding paste into an empty fixture nest, solder paste application lends itself quite well to automatic assembly.

The use of controlled atmospheres and the gradual rise of component temperature to a specific desired point make brazing a most reliable means of joining. It has a somewhat limited application to those products that can stand the brazing oven temperatures without damage and where surface finishes are either not critical or will be subsequently plated or coated. Since the joining procedure does not actually occur on the assembly machine, which only prepares the parts for assembly, joining times and line imbalance are not significant. Depending on the physical integrity of the assembly prior to ejection to the oven conveyor, parts may have to be placed in carbon boats or other pallets or carriers to go through the brazing oven.

Soldering on an assembly machine poses another type of problem. Fluxing materials may be (and usually are) required in lieu of a controlled atmosphere. Flux is corrosive, and if not carefully controlled, may accumulate or form a residue on fixtures, conveyors, and machine stations. It also means that assembled parts may have to be cleaned or washed prior to any additional operations or packaging. In order to get around this problem of flux contamination and the associated problems of flux dispensing, one might consider utilizing center-core solder having the fluxing material within the tubular like solder material, or using pretinned or flux-coated components. Because of the unique requirements of each different assembly, a great deal of concept research will be necessary to find the best approach.

Some years ago, the use of ultrasonic excitation of a solder pot seemed to offer great possibilities for those solder operations where parts may be dipped into molten solder. The underlying theory was that the motion of the solder, energized by an ultrasonic horn, would scrub away oxidation on the surfaces to be bonded, thus eliminating the need for fluxes. Horn deterioration through etching and dross accumulation proved troublesome, and this approach is no longer used.

Assembled parts may also be ejected to fountain soldering units, soldering baths, or a newly emerging approach, vapor-phase soldering. Vapor-phase soldering requires that the assembly system deposit solder paste in the appropriate amounts, precisely located, and verify the presence of this paste in some way, possibly through optical scanning.

A second area of concern when attempting to solder on any assembly machine is that of providing adequate heat transfer in those areas of the assembly where the heat is required. This problem of heat transfer is significant since heat transfer requires time, and time is a precious commodity on assembly machines. The transfer of heat can be done through physical contact of one or more parts to be soldered to a heated element (or simply a soldering iron), by exposing the parts to be assembled to a gas flame or by positioning them in the proximity of a calrod unit or specialized heat jets or lamps.

This heat transfer is a function of time and must be adjusted to the index and dwell cycles of intermittently indexed machines. When the machine is in motion and continuously operating, there is little problem. The problem comes when the machine is stopped and the parts (and their work-holding fixtures) begin to cook away in the heat. Some machines have overcome this problem by using a continuous index, but this increases the complexity of part transfer stations and inspection probes.

The use of specially shaped induction coils to transfer heat in the precise amounts and in the precise desired location has proven successful in some applications. Coil design is usually empirically developed. It may be necessary to incorporate ferrite cores in the coil. Once a coil form is established it may be necessary to epoxy coat the coil assembly to prevent accidental distortion of the soft copper coil material. Cooling water may be required, often meaning the use of distilled water and an associated refrigerant system. Fixture designers should be aware of the microscopic erosion of metal fixtures in an induction field. All this must be balanced against the enormous advantage of being able to apply heat precisely when and where required and trigger that heat application only when conditions are proper. Soldering is one of the most difficult joining procedures in automatic assembly but it is also potentially one of the most profitable areas of possible mechanization.

Environmental concerns about lead poisoning are causing major process changes, particularly where food or liquids can come in contact with solder, but also where workplace contamination is occurring.

JOINING WITH HEAT AND FRICTION

Welding

Resistance Welding. Resistance welding is particularly well suited to automatic assembly. For this joining process, there is generally no need to introduce material such as fluxes or solders, but rather the materials are joined by fusion under controlled pressure when extremely high currents are passed through the parts. Environmental restrictions on coatings to facilitate welding are of concern.

By virtue of the mechanical control and repeatability requirements that are associated with various transfer and inserting operations, automatic assembly machinery are well adapted to carrying electrode holders. As a result of their consistency of performance these machines have produced drastic reductions in the scrap that usually results from improperly positioned parts or improper electrode follow-through during welding.

Pressure, current, and frequency relationships for welding operations have received a great deal of attention. For any combination of materials, the best relationship of these variables must be determined for each application. When welding operations are included in assembly machines, certain basic questions must be resolved before the best welding method can be determined. These questions are:

- What is the actual weld time and hold time?
- What is the weld pressure required?
- What is the amount of material collapse that might be expected as a result of the welding operation?

Most transferral operations on assembly machines are practically instantaneous. Therefore, production rates on such equipment are governed by the feeder efficiency and the mechanical and inertial limitations of the machine system. However, in welding machines additional limitations also are imposed by the nature of the weld-

ing operation. Although welding is relatively rapid, there are certain specific time requirements. Consider the total cycle for an assembly machine. This cycle will consist of the time required for index, the time for lowering and raising of the electrodes to apply proper pressure, the time to apply the welding current, the holding time necessary under electrode pressure to allow the weld to cool, and, finally, the time for removal of the electrodes prior to index. Machine rates of 40–90 strokes per minute are typical of that which may be obtained with automatic welding equipment. However, the actual welding time for each specific assembly must be calculated prior to determining assembly machine production rates. The machine shown in Figure 7-17 is capable of welding 3600 assemblies per hour.

Uniform pressure application during the welding operation is essential for good welds. As the material at the interface reaches a molten state during the welding cycle, the individual parts of

Fig. 7-17. A rotary assembly and welding machine uses cams for the raising and lowering of welding electrodes.

the assembly can move toward each other causing a reduction in length or height in both the parts and the assembly. Welding pressure will drop off during the welding cycle itself if the electrodes do not have the ability to follow-through and to maintain the required pressure throughout the weld. The mounting of electrode carriers in antifriction bearings facilitates this follow-through which is an essential requirement. The applied pressure is critical to creating sound welds from a metallurgical standpoint. The need for ensuring free movement of the electrode holders without restriction from the connections to the welding control station or water coolant lines cannot be overemphasized. It has been found that a preloaded upper electrode holder, using the proper series-parallel arrangement of Belleville-type spring washers, permits close control of the application of pressure through the upper electrode.

To raise the lower electrode to the work and to have it still able to resist the downward pressure of the upper electrode, a toggle arrangement is possible on any cam-controlled assembly machine. This type of toggle linkage can withstand high pressures, and is also capable of fairly long withdrawal motion of the electrode carrier prior to index. Lighter resistance welding applications are shown in Figure 7-18.

Because of the high welding rates possible through automatic indexing of the workpieces under the welding head, extremely high operating temperatures in both the control unit and the electrodes can be produced. It may be necessary to use water as a cooling medium to keep this heat within reasonable limits. Cold water, whether obtained from a water main or a recirculating cooling system, can be expensive. Freestanding refrigerant systems for cooling water are becoming increasingly popular so that water is circulated only when required.

Welding fixture designs must take into account the problem of heat buildup. The use of normal tool steel for fixture nests is not usually wise. Stainless steel is often used, as well as Micarta, other plastics, and copper alloys.

The presence of high welding currents presents two special problems for assembly machines: the transmission of heat to the machine and the de-

Fig. 7-18. Light resistance welding is done on this linear machine. Electrodes are carried in low inertia holders and weld quality is monitored through toroidal coils measuring field strength.

velopment of induced magnetism or improper current flow. Generally, proper fixture design and the use of suitable materials in the workholding fixtures and indexing dial can greatly alleviate these problems (see Figure 7-19).

Indexing dials or pallets made of laminated phenolic material have proved extremely heat stable and are well suited to machining and jig boring operations. The use of hardened thread inserts eliminates the tendency for bolts holding the work-holding fixtures to strip this material. Also, the high insulating qualities of the laminated phenolic material prevent erroneous current flow to occur and, in effect, isolate each work fixture from the rest of the machine.

Certain copper alloys have been developed for workholding fixtures in welding equipment. These alloys have a low resistance to current flow, and can be used for current shunts when the design of the assembly components makes this necessary. Because they are nonferrous, the alloys do

Fig. 7-19. Welding can be very precisely controlled on automatic assembly systems.

not tend to become magnetized. In addition, they have excellent heat-dissipating qualities. In some instances, care must be taken to ensure that the loops formed by the cables connecting the electrode carriers to the welding control unit do not become inductive coils and set up lines of magnetic flux.

Electrodes must be considered as perishable. When they are integral to the work fixture as lower electrodes, they must be replaced when they become worn to the point that accurate location of the assembled parts is jeopardized. Experience has proved that molybdenum-faced lower electrodes greatly increase electrode life.

Resistance welding is an excellent joining procedure for incorporation on automatic assembly. The major design problems are electrode wear,

electrode changeover and setting, fixture wear, fixture impact on the welding field, and the monitoring of weld performance at each stroke of the machine. The first four of these problems are those of tool design and experimentation. The last, the monitoring of weld quality, lends itself readily to available commercial units. One such monitor shown in Figure 7-20 measures weld field strength by use of a toroidal coil. It can then determine which welds fall outside empirically developed parameters and segregate parts, or shut down the machine to facilitate examination or replacement of electrodes.

Inert Gas and Plasma Welding. Inert gas welding such as TIG and MIG welding lends itself to automatic assembly. Plasma welding has also

Fig. 7-20. A weld-monitoring device determines whether the weld fell in empirically established limits. Upper and lower limits are set on digital switches while a display illustrates the exact character of the last weld.

been successfully coupled to assembly systems. Figure 7-21 shows a TIG welding unit used to join lead components.

Laser Welding. Lasers have been integrated with ease into automatic assembly systems. They offer high utilization rates, since there are no electrodes to wear or deteriorate. Design problems center around proper guarding of the laser path and fixture damage that might occur should no parts be present when the laser welding occurs.

Laser welding requires more precise control than laser drilling since the parts must be heated but not allowed to liquefy or vaporize.

Ultrasonic Welding

Ultrasonic welding is the joining of two parts together by the friction and heat generated

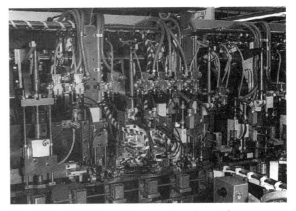

Fig. 7-21. A TIG welding system is used to secure weld two lead components together. The head must approach the work, initiate the weld, and traverse in a short period of time.

through minutely rubbing them against one another at approximately 20,000 cycles per second. This rubbing action causes the surface molecules of both parts to intermingle before they resolidify, so there is essentially a molecular bonding of two materials.

Ultrasonic welding techniques can be used to join dissimilar metals as well as metals that are conventionally welded. It finds broad application in the joining of thermoforming plastics. It cannot normally be used with thermosetting plastics such as bakelite. Ultrasonic energy can be used to deform parts, but competes with heat staking technology in this application. Since it is normally intended for surface bonding, it can only deform parts by introducing large amounts of energy to the part piece from the ultrasonic horn. In some instances introduction of such high lev-

Fig. 7-22. Ultrasonic welders are used to join plastic parts. When mounted on assembly machines, great care should be taken to ensure positive withdrawal prior to index.

els of energy causes the material to squirt away from the horn in an uncontrolled fashion.

When attempting to upset plastic materials having some type of physical memory, it is worthwhile experimenting to see if gross deformation can be done first by pressure and ultrasonic energy then applied to modify the part piece memory to the new configuration of the part.

From a machine design standpoint, it pays to think of ultrasonic applications as an audio problem, for it involves excitation or vibration of parts in a controlled fashion at a frequency level within the audible range of many humans.

Effective ultrasonic welding depends in great part on components designed for efficient weld-

ing. This has been discussed in Chapter 2. The provision of energy concentration features can make a difficult job easy, but some experimentation may be required to determine the proper part configuration.

Efficient horn coupling is the next significant factor. The shaping of the horn not only to properly contact the part piece but also to resonate properly requires skill and practice. Titanium horns are common, and it may be necessary to try several horn shapes before being satisfied. This horn development activity should be considered as a necessary development cost. (Figure 7-22.)

Fixture design is another factor for effective ul-

Fig. 7-23. A cam-actuated override device positively pushes the welding unit upward immediately prior to index to prevent horn damage.

trasonic welding. Proper staging of the parts without undue sympathetic vibration is most important. It is essential to do horn development with parts being held in the proposed production fixture. The combination of horn, parts, and fixture complete the acoustical system.

From an operational standpoint one must recognize the comparative fragility of both the welding horn and piezoelectric transducer. A standby horn should be readily available. Most ultrasonic units advance the horn to the work by use of air pressure.

During welding, follow-through occurs automatically before withdrawal as air in the cylinder is reversed. Because of the sensitivity of the unit to side thrust and accidental damage, it is wise to insist that the machine have some form of either positive withdrawal prior to index or some absolute form of interlocking. Figure 7-23 illustrates such a system.

Heat Staking

The use of heat staking is still common in assembly. In one extreme form, hot upsetting of metal, electrical current from a resistance welder is used to heat metal to the point that it becomes malleable under electrode pressure.

Ultrasonic energy is another way of developing deformation. But the application of simple heat is often enough to cause sufficient malleability to deform a part to secure an assembly. The basic problems with heat staking are those of closely controlled temperature and the tendency of parts to stick to the heating element. Silicon coating of the element has proved helpful in many applications.

SUMMARY

Almost every conceivable joining procedure is applicable to automatic assembly. Each has its own application problems.

The efficiency of the joining process will be in great part determined by product design and part quality. Each of these processes will require more attention to safety guarding than will be the case in part feeding and transfer.

The use of radio frequencies in many welders, induction units, and similar power sources may present some hazard to those wearing electronic pacemakers. To date the extent of the hazard is unknown, but warning labels should be affixed to such equipment indicating potential hazards.

Chapter 8
Inspection, Gauging, and Functional Testing

INTRODUCTION

Social values and expectations have placed new burdens on quality assurance managers. Statistical sampling techniques have no room where consumerism, government regulations, and international competition mandate total quality control.

Automatic assembly machinery has a unique capability, if recognized, of guaranteeing that 100% of all assemblies have all of the necessary components, that these components are in the proper position, and that the assembled product is functional—all of this at modest cost.

Other chapters have discussed quality considerations as they relate to incoming part quality levels, machine productivity, and control system selection. This chapter will cover the types of sensors used to detect machine tooling and operator safety, component part presence and position, and the types and limitations of functionally testing assembled products on the assembly machine. Since many testing procedures require extended periods of time, the use of functional testing can have significant impact on machine system configuration, overall costs, and machine productivity. The economic and technical feasibility of *functional testing* on the assembly ma-

chine must be determined early in machine layout and development.

This chapter will discuss the types of sensory devices available for use today and the coupling of their signals to the machine control system. It will also discuss an area of enormous corporate concern—manufacturing quality documentation. Product liability lawsuits, class action lawsuits, and mandatory product recalls are placing new emphasis on serialization, date coding, and recording of quality characteristics as a means of eliminating or limiting product liability exposure. While these coding and recording devices are not quality sensors, their use is directly related to the topic of this chapter. Customer perception of product quality has become the dominant marketing factor. Modern assembly system design can ensure that perception is based on reality, reducing warranty expenses and product recalls.

INSPECTION FOR WORK ENVIRONMENT SAFETY

In the order of human priorities a safe working environment has taken a high place. Machine safety and operator safety must be part of the system procurement and initial design. In some hazardous work areas, barrier guarding is necessary. In other areas of assembly mechanization the necessity for frequent access is such that inspection-like devices are necessary to monitor the presence of operating or maintenance personnel in the area.

Major capital investment such as an integrated automatic assembly system often represents for a corporation the sole global source for the production of a critical product line. Major system damage that would put a large machine out of production for extended periods must be avoided at all costs. A variety of sensors and switches may be used to detect conditions that could lead to severe machine damage; these include air pressure switches shown in Figure 8-1, temperature sensors, and interlocking switches which indicate the location of machine elements.

Inspection units to ensure operator and machine safety are not normally thought of as inspection units, but in their operation they duplicate those inspection units which monitor product quality. These output signals are fed to a machine control system in the same way as those used to ensure quality of production.

There is one essential difference in that most inspection stations that shut down the machine for reasons of safety are usually designed to pre-

Fig. 8-2. Braking speeds on mechanically controlled machines can be adjusted to the need with this controller. Emergency stops are rapid, but stops for product quality reasons are more gentle.

vent any powered operation of the machine until the problem is corrected. Quality inspection stations usually are so wired that the machine can be jogged or inched under power to facilitate correction of the problem, or the system can continue production while segregating defects.

When machines must be stopped on detecting a machine or operator safety problem, braking action should occur as quickly as possible. When machines are stopped for quality assurance problems, machine stoppage can be done somewhat more leisurely or even postponed momentarily until tooling is withdrawn. The brake control shown in Figure 8-2 is designed to provide rapid braking when required and gentler braking action when time is not critical.

INSPECTION FOR PRODUCT QUALITY

Inspection units on assembly systems which monitor the developing quality of a product at each step of the assembly process are designed to pass along without any corrective action those assemblies that give indication of being properly assembled or have required functionality up to the stage of assembly then being monitored. They are also designed to identify those products that lack some desired quality or characteristic which the product should have at this point in the assembly

Fig. 8-1. An air pressure switch monitors air pressure and shuts the machine down automatically before improper pressure could cause machine damage.

process. Once identified, the machine's control system can cause one of several actions. The machine can be programmed to stop instantly for manual correction of the problem or transfer the defective assembly to a manual repair loop before reintroduction to the main assembly system. The machine can be designed to eject the defective part immediately after detection of the problem. The machine can be designed to carry the defective part through subsequent work stations that will be locked out of any further work on that assembly until the defective assembly is segregated at ejection. The machine can be programmed to pass the assembly completely through the system back to the station in which the problem occurred. Lastly, the machine can be designed with redundant stations that will make a second attempt to complete the failed operation after detection of the first.

Quality monitoring stations on assembly machines will be basically one of two fundamental types. The first will inspect for failure of the assembly process, while the second will verify, where possible, the functionality of the assembly up to that point of completion. These two types are essentially different. One monitors the machine's success in performing the assembly task. The other determines if the physically successful assembly operation has in fact produced a product that is also functionally acceptable.

The most common type of inspection unit found on any machine is one that verifies the presence and position of the product component or fastener fed or inserted in the last previous station. A typical station is illustrated in Figure 8-3.

Many early assembly machines did not have inspection units for presence and position of each assembled component. Instead they depended on process reliability. These earliest machines were usually designed for products manufactured in very high volume with extremely long product or model life. These very high volume, long product life applications usually are best handled by a series of prototype machines and small adjustments to component design leading up to very high process reliability at assembly.

The thrust of most present assembly system projects is oriented toward building a one-of-a-kind machine for more modest production volumes and limited product life in the shortest practical time.

Fig. 8-3. Inspection for presence and position of a component part is done by this dual switch inspection unit. If no part is present, the probe will not compensate upward and the lower normally open switch will not be closed. If this part is not fully inspected, excessive compensation of the probe will open the normally closed upper switch.

There usually is too short a lead time and such financial constraints that prototype development of such a machine is not practical nor is dependence on process reliability truly feasible for today's insistence on quality assurance. Today's political-social climate is unwilling to accept random failure (no matter how rare) without penalty. Verification of assembly quality is essential. This verification process will include three aspects: quality of the incoming components, quality of the machine's assembly and joining functions, and quality (functionality within design limits) of the assembled product.

Before examining these three aspects, however,

one must recognize an often overlooked point in any discussion of inspection, gauging, and functional testing on assembly machinery. Most inspection stations on assembly machinery do not identify the cause of a problem but only its existence. If a gauging station shows a part not fully inserted, it does not usually identify the problem an oversized part, an undersized hole in the receiver, a feeder failure to present the part properly for pick up, a transfer unit failure to insert the part fully, or misalignment of the indexing fixture to the fixed transfer unit. If a leak testing station indicates pressure or vacuum deterioration, is there a defect in the housing or cover, a tear in a diaphragm or seal, failure to join hermetically, or improper engagement of the leak testing unit to the product? The average inspection or test station does not have the capability of assigning causes but merely identifies the problem, its frequency, and, in a well designed system, the impact in terms of downtime that this problem has on machine productivity.

In essence, most assembly machine test stations perform a quality control function but not a quality assurance function, if one takes the meaning of quality assurance in its proper sense. If quality control ensures that the customer does not receive a bad product, quality assurance attempts to ensure that the manufacturer does not build a bad product. Identified assembly problems must be analyzed off machine as to the source or nature of the problem.

Quality of Incoming Parts

An assembly machine, designed and built without any inspection units, will be in its very operation a fairly good parts sorter. Oversized or incomplete components will jam feeders, tracks, and fixture nests. Foreign matter and manufacturing scrap such as sprues, flash, or chips will cause similar problems. Most manufacturers with extensive manual assembly lines are stunned when they attempt to start up their first assembly machine to find the degree of contamination and quality levels of parts coming to the assembly line. One is forced to recognize, sometimes for the first time, how much of the cost in a manual assembly line is incurred in segregating defective parts, and in the assembly of part pieces not meeting print specifications. Figure 8-4 illustrates this point.

Fig. 8-4. A sorting feeder processing two boxes of threaded fasteners found this selection of defective parts, foreign parts, and scrap.

Unfortunately, corrective action on this problem usually follows machine installation, since the extent of the problem is usually not known at time of fixture and feeder development. This is especially true in concurrent engineering projects. Since a very high proportion of present-day assembly machines are built for a new product, concurrently with development of fabricating equipment, the full extent of the problem may not be known until full production operations are tried. Usually, however, an analysis of existing production of similar products will give a good indication of probable incoming part quality levels.

Once the extent of the problem is recognized, corrective action will take one or more of four paths. The first two are worthwhile, one is quite costly, and one usually useless.

The first corrective action and the most productive is an upgrading by improved manufacturing and deburring processes, by better housekeeping, and by improved quality control of the components brought to the assembly machine.

This is best accomplished if manufacturing and assembly are integrated under one plant management. All too often, however, assembly areas are located at some distance from part manufacturing operations. Improving incoming part quality through improved manufacturing discipline will add cost that must be justified through further improvement in assembly productivity. There is no such thing as a free lunch.

Breakdown of incoming part quality will usually occur when purchasing has failed to adequately specify and obtain guarantees of part quality, or when rejection of an incoming lot of parts will shut down the machine and production requirements mandate its operation.

The second corrective action is to presort in sorting feeders parts that will subsequently be placed in the assembly machine feeders. Presorting hoppers will serve a useful task if the incidence of physical part defects is high. Usually, the need for sorting feeders will become apparent at time of hopper development, but this usually will only be detected by the user rather than the feeder manufacturer. This subject was discussed in greater detail in Chapter 4.

The third corrective action once the problems with part quality are recognized is an attempt to modify the machine to accept a broader range of nonfunctional part tolerances than was anticipated from examination of product drawings. A certain amount of this work is normal during the debugging process and is clearly beneficial. There comes a point in this accommodation process where there are diminishing returns.

Opening up tracks, fingers, and fixtures to accommodate larger and larger parts begins to increase the number of failures when attempting to feed, transfer, or locate significantly smaller parts. Once this happens overall machine productivity drops dramatically and can be corrected only by major reworking of the tooling on the machine.

These first three corrective actions do not require gauging or inspection stations per se. The fourth often suggested corrective action does. Many builders are often told by customers that a high incidence of bad components may be expected (for example, 3% or 4%) and, hence, it will be necessary to inspect the incoming components for quality as part of the assembly machine func-

tion. This suggestion, if accepted by the builder, will usually lead to severe problems, even insolvable problems, at time of acceptance runs, extended debugging, and ill feeling between builder and customer.

For instance, if one intends to do incoming part quality monitoring in the fixture, one must be prepared to use a memory-mode type control system, because it is not economically feasible to stop the machine every 30 or 40 strokes to remove a bad part. There are enough problems with machine-related failure and there are many stations each with their own problems.

Others suggest that some type of parts quality inspection be done in the feeder tracks with alternative ejection of defective parts done in the feeder rail. The feeder rail now becomes a small gauging machine and must have adequate reservoirs of parts before the gauging station and after the gauging station. Any successive rejects will generally starve the transfer unit and cause frequent machine stoppage.

The use of an expensive assembly system to do part sorting is usually a most impractical approach.

Quality of Machine Functions

The primary role of inspection units to verify presence and position of components on assembly machinery is to monitor the actions of the assembly machine where insufficient confidence exists in process reliability. The incorporation of such inspection stations, such as shown in Figure 8-5, add a minor amount to the overall systems cost. Verification of the complete transfer and inspection of each component as shown in Figure 8-3 is the most common type of inspection. On assembly machinery manufactured in the United States, such inspection is usually done with independent stations. In Europe and Japan, perhaps because of the heavy use of rotary machines, and because of severe constraints in available floor space, one often sees an attempt to incorporate sensors in the transfer fingers or jaws rather than dedicate individual stations on the machine to inspection roles. One also sees attempts at robotic assembly incorporating tactile or pressure sensors in the gripper jaws. While these sensors may indicate the presence of a part during the transfer

Fig. 8-5. Inspection units located after each transfer station verify the presence and position of each part piece in the assembly sequence.

operation, they do not adequately inspect for the completion of parts insertion.

While insertion of each component is easily verified, any inspection for the success of joining operations is much more difficult. One must be prepared to accept some reliance on process reliability determined by off-line testing. Torque testing, push out tests, and tensile forces can be incorporated into test units, while other stations can monitor deformation of rivets, etc., but one often will find that it is necessary to seek some type of implicit rather than explicit verification that the joining procedure did occur. In some instances one can only look to the functionality of the product to verify that the joining process was completed properly.

Quality of the Assembled Product

Even when the assembly machine has performed all of its operations properly and successfully joined the components to form an integral product or subassembly, the result may not be satisfactory, for the assembled product may lack certain intrinsic physical properties or functional characteristics.

Inspection stations designed to determine the quality of the assembled products may be incorporated into the machine system just prior to ejection of the completed product but also at intermediate stages of assembly and often before final joining or sealing operation where salvage of defective assembly might be made more difficult or impractical.

Dimensions of Assembled Parts. One of the real challenges of any assembled product is to ensure that the buildup of the accumulated tolerances in each component part results in an assembled product with overall physical dimensions that are functional. Incorporation of dimensional gauges on assembly machinery can be of any degree of complexity. Relatively large parts can be detected by a unit as shown in Figure 8-3, but these, when used to monitor product dimensional characteristics, are limited to gross dimensions in excess of 0.020 in. (0.5 mm). For closer dimensional monitoring (or to verify the presence and position of thin parts) a more precise inspection device may be required, the gauges shown in Figures 8-6 and 8-7 will prove useful.

Fig. 8-6. A precision snap-action switch will measure to close limits and can be adjusted to precise tolerances, often within 0.001 in. (0.025 mm).

When using switches of this type it should be remembered that measurement will be unilateral, that is, a switch-actuated device can only measure one limit; for instance, is the part at least 0.025 in. (0.6 mm) thick? It would require two switch-actuated devices to monitor minimum and maximum acceptable limits, such as a part must be at least 0.025 in. (0.6 mm) but no more

than 0.032 in. (0.8 mm). When ranges are required, the use of a transducer type of probe feeding to a control unit that will provide an isolated output signal only when parts fall within a specific range may prove more efficient. A typical example is shown in Figure 8-8.

Any attempt at dimensional gauging will be dependent on the ability of the gauge to have some point of reference. Either the fixture or, more hopefully, some assembly component will have some staging or reference surface for accurate reference to the dimensional characteristic being measured. This requirement was earlier mentioned in the chapter on product design for efficient assembly.

Some components falling in the overall acceptable design limits may require further segregation into various smaller increments within the overall range for purposes of matched assembly. Transducer output is passed through a series of band-pass filters to specify or control further assembly action. A typical case would be shown in Figure 8-9. A spring-loaded hydraulic valve is preassembled and its movement measured under a specifically preloaded transducer. Once the

Fig. 8-7. When very precise measurements are required, small continuity gauges connected to intrinsically safe relays permit measurement to within 0.0005 in. (0.013 mm).

Fig. 8-8. An air gauge is used to ensure that an assembled bearing is within acceptable upper and lower limits. While the gauge indicator shows actual dimensions on each part, the machine's control system receives a binary signal, indicating whether the part fell within the acceptable range without any reference to the actual dimension.

Fig. 8-9. A classifier such as that shown here is used to segregate measured parts not only into below minimums, acceptable, and above minimums but to separate acceptable assemblies into small range limits.

movement is determined (in effect a buildup of part tolerances and spring characteristics) the classifier will then indicate which spacer washers should be inserted to bring the valve within specific operating tolerance.

Occasionally one may find that product acceptance may require that two dimensions must be measured against one another and both against a third surface. The use of two classifiers feeding

into a Wheatstone-bridge-type circuit can be used successfully as illustrated in Figure 8-10.

Quantitative Measurement and Analysis. Some products may not only require a determination that they fall within a specific static range, but that they have certain dynamic qualities. For example, a rheostat may have to show specific resistance characteristics at certain degrees of ro-

Fig. 8-10. Two transducer signals were fed to controllers connected in Wheatstone bridge fashion, when it was necessary to measure two operating rods which could protrude through a third part within a tolerance of 0.020 in. (0.5 mm) but could not differ one from the other within 0.010 in. (0.25 mm).

tor rotation. Such stations, therefore, need not only accurate coupling to sensors but also controlled mechanical movement coupled to sensor monitoring. The widespread use of programmable controllers, having motion control, timing, and counting capabilities, has opened up broad possibilities in such dynamic testing procedures.

Calibration. Closely allied to quantitative measurement is the field of product calibration. Many products require some adjustment after assembly. Solid-state electronic devices may require laser trimming of substrates. Relays and circuit breakers may require physical deformation of reeds or the adjustment of stops under known electrical or magnetic forces. Here mechanical motion may not only have to be coordinated with measuring sensors but will actually have to be controlled by the sensors. Sometimes the assembly system will be required to do pre-calibration, that is, rough calibration bringing the assembly into the range of final calibration equipment without hunting.

Each of these four types of product quality

measurement is increasingly more complex in the demands it places upon machine and control design. In order for any of these measurements to be incorporated into automatic assembly machinery, the monitoring equipment must be suitable for high cyclic operations in the manufacturing environment.

INSPECTION DEVICES

Before examining the general categories of inspection devices it should be noted that any of these proposed devices must be capable of being used repetitively in a high-volume manufacturing environment, if they are not to lower productivity. The use of inspection devices coupled to the assembly machine should not be a source of machine downtime.

Criteria for the Use of Inspection Units in Mass Production

The coupling of any inspection unit to the product being inspected is a task for the assembly system designer. Answers to unique problems relative to machine access, product configuration, and necessary isolation from vibration, noise, electrical fields, etc., can only be handled on a case by case basis. There are, however, certain criteria that will determine if a proposed inspection technology is practical on an automatic assembly system.

Response Time. In any mass-production type operation, line balancing is a critical design element. Automatic assembly done on synchronous machines will have its cyclic rates determined by the single longest operation on the machine. Since most insertion work is almost instantaneous, insertion is rarely the limiting factor in assembly system productivity, and whatever component insertion constraints occur are due to feeder limitations and inertial considerations in the basic machine design. Joining operations usually will require a greater portion of total cycle time and mean longer dwell time requirements to complete necessary operations. The most inhibiting operations in mechanized assembly are usually inspection functions. While presence and position inspection units are like part insertion in that they require almost no significant time, other inspection operations may require the stabilized establishment of a condition or physical environment before any measurement can be taken. Cavities may have to be evacuated before pressure or vacuum decay may be tested. Parts may have to be raised to certain temperatures before testing can occur. Back pressures may have to be established for air gauging.

Even inspection probe advance to the workpiece may have to be constrained. Too rapid an advance may cause bounce or vibration on lightly loaded units causing erratic signals until the probe is stable. Production systems operating above 50 cycles per minute will be inhibited by inspection device constraints. Accurate determination of cycle times, including sensor advance, establishment of required conditions, measurement of characteristics, withdrawal of the probes, and index time will often establish system cyclic rate.

It may prove that the restraints on inspection

cyclic rates are totally incompatible with production requirements for the assembly machine. If this testing is done after all assembly operations are complete, completed assemblies may be ejected from the assembly machine to a battery of test stands. If the test requirements are necessary during the middle of the assembly process, and test cycle time is incompatible with required machine rates, it may be necessary to send the parts off the machine and then return them after testing.

Both synchronous and nonsynchronous machines are capable queuing techniques or parallel paths to perform line balancing requirements for long inspection procedures, providing the inbalance between assembly requirements and testing requirements does not exceed a three-to-one ratio.

Fail-Safe Operation. In the integration of any inspection station on an assembly machine, the integration of that monitor to the machine control system and the coupling of the inspection probe to the part must be such that *any failure of the probe, control, or coupling must indicate part failure*. In an age acclimated to the acceptance of electronic displays and computer printouts, one can be lulled into acceptance of erroneous signals. Many ordnance plants, which have unique quality and safety requirements, use a system of quality logs and master products to sign off each hour's production. At each hour an inspector sends a test master through each inspection probe. Failure of any inspection unit to respond accurately to the master piece is reason to institute reinspection of the previous hour's production.

Operating Life. In assessing the feasibility of incorporating a proposed inspection gauge into an automatic assembly system, the durability of the proposed gauge in a high-production operating environment must be established. If necessary, redundant stations should be provided for. Installation of sensitive equipment should be free from vibration and excessive heat buildup and high humidity. Standby electronic test equipment may be maintained in better operational condition when power is provided to the alternative control in a standby mode.

Probes like all tooling on an indexing assembly machine should advance to the workface under the lightest possible preload consistent with avoiding bounce or flutter, but should retract from the assembly or component under test in a positive manner prior to index.

Inspection by Simulation and Equivalence. Test procedures should not in themselves be harmful to the product, the operator, or the machine. For that reason test procedures that lend themselves to the laboratory may not prove practical on the assembly machine. In many cases *simulation of the test procedure or substitution of an equivalent procedure may have to be considered.* The suggestion of such simulation or substitution will probably meet initial resistance from those who created the original test procedures. The not-invented-here syndrome is found everywhere. In fact, simulation or equivalence may prove to be the only means of automatic inspection at production rates and may have to be backed up with statistically based sampling on a laboratory basis.

Types of Inspection Devices

The means by which product characteristics can be measured is almost infinite. The majority of required operations will, however, be done by one of the procedures tested below.

Mechanically Actuated Switches. A switch is nothing more than a device that opens or closes a circuit. It is hoped that it will do so without pitting the contacts of the switch at each opening or closing. The mechanical snap-action limit switch, often generically if erroneously called a microswitch, is a basic tool in many inspection devices. Probes that are preloaded in one direction will either open or close certain switches by the amount of axial travel of the probe in its housing caused by contact (or lack of contact) with the workpiece. Figures 8-3 and 8-6 illustrate such switches. Figure 8-7 illustrates another type of switch where snap-action characteristics are not essential, because the current

on the switch mechanism is limited by an intrinsically safe relay, hence pitting of contacting elements is eliminated. Mounting of switches should be in accordance with good machine tool practice.

Proximity and Magnetic Devices. No matter how well constructed, limit switches have finite lives and are subject to mechanical wear. Proximity switches sense the movement of probes through changes in a magnetic field induced by the presence of an object within that field. Control complexity is increased, for the simple opening and closing of the circuit by a switch is now done through a relay. Proximity devices can be actuated by probes which are in contact with the workpiece or, in some instances, by the proximity of the workpiece.

Capacitance and Inductance. Part sensing can also be done by changes in capacitance or inductance which occur because of changes in the relationship of a moving element to a fixed element. These changes can be in the form of an analog output rather than a binary or digital output. The classifying gauge shown in Figure 8-9 depends on minute analog changes which occur when the moving part of the transducer changes relative to the fixed portion of the transducer. These minute changes can be amplified to detect extremely small dimensional changes.

Air Gauging. Air gauging depends on the ability to sense minute changes in air pressure which result when resistance to air flow in gage orifices changes. Air gauging is particularly adapted to the measurement of cylindrical parts or holes. Gauges such as those shown in Figure 8-8 require a source of clean dry compressed air. They automatically compensate for axial dislocation of the gage to the part being measured. They require a relatively long cycle time since air flow must be stabilized once the gauge is placed in proximity to the surface being monitored, before an accurate reading can be taken. This inhibition has restricted the use of air gauging in high-speed machines, but its accuracy of measurement, its elimination of the need to physically control the workpieces, and its lack of sensitivity to axial displacement make air gauging well worth considering.

In a more gross fashion, air streams or jets can be used to monitor the level of parts in a transfer rail very effectively since their slow response time provides a built in time delay relay which is necessary for the use of photoelectric cells that would give erroneous signals during normal passage of parts in transfer tracks.

The entire field of pressure differentials seems to be overlooked as a means for accurate noncontacting assembly gauging and inspection.

Fig. 8-11. A "leak tester" shown here can monitor microscopic deterioration of pressure by comparing the test item to deterioration in a known orifice or a known chamber or to the actual pressures in the tested pieces at two different times.

Pressure and Vacuum Testing. If air gauging is the measurement of the degree of restriction to air flow (and hence an analog signal) by means of pressure differential measurement, then the ability of an assembly to maintain a pressure (or sustain a vacuum) can also be so measured.

The current social importance of containing pollution and ensuring fuel economy has placed great emphasis on products that must establish a positive pressure or vacuum for effective operation. The common name for such detection of pressure or vacuum deterioration is "leak testing." In a gross sense leak testing can be done by placing a pressurized body under water and looking for bubbles or by coating a part with a soapy solution and looking for bubbles. Most of us cannot find the time to experience joys of "always blowing bubbles." More microscopic leaks can be detected by utilizing gas detection devices to sense parts which were pressurized with tracer gases. This technique, called "sniffing," works well when only microscopic leaks occur, but it is a difficult technique for automation since assemblies which have gross leakage will overwhelm the "sniffer" and the equipment must be shut down until the sensors regain equilibrium.

For high-production assembly machines there are three possible techniques: flow type, differential pressure decay type, and electronic memory pressure decay type. In a flow-type unit differential pressure between a known restriction and the item being tested is measured. In the conventional pressure-decay type pressure in a chamber is compared to the item being tested. (See Figure 8-11.) In a memory-type unit initial test pressure of the stabilization is compared to a test pressure after a programmed time period. In each case actual testing can be done only after stabilization and measurement of decay. These requirements take finite time periods. If testing is required at rates much exceeding 15–20 tests per minute, redundant stations may be required.

Several observations are worth noting. There is no such thing as "no leakage is allowed." Everything leaks. It is a question as to what rate of leakage is permissible. Leak testing will depend for its accuracy on proper calibration and a proper sealing of the workpiece to the gauge.

Sealing materials that work properly may need frequent replacement. One should also be aware of the effect of pressurization on the air mass used for testing. As this air is compressed it will heat up and expand. During stabilization it will cool off and shrink. This is not leakage. There have been tremendous improvements in the stabilization and response times of "leak testing" equipment. Suppliers have worked diligently to reduce total inspection times equivalent to machine cycle requirements.

Photoelectric Devices. Photoelectric cells can be used to determine the presence of a direct or reflected beam of light. When reflected, they indicate the presence of a part, and when direct, the beam indicates the absence of a part. Their use in automatic assembly is varied. They are often used as in Figure 8-12 to monitor track levels. Electric eyes suffer in the field from vibration, misalignment, and accumulation of dirt on either of the lenses. While their versatility is quite high, their durability is not.

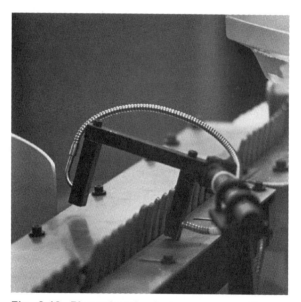

Fig. 8-12. Photoelectric devices measure the reception of a known light source or its interruption. They are often used to monitor the level of parts in a transfer track as shown here. When the track is full, the main feeder bowl shuts down, reducing noise and wear.

Vision Sensors. One of the most interesting areas of inspection is that of object recognition by some form of visual sensor. Many are familiar with one form of such equipment in supermarkets that use the product codes on each sales item (a series of different width black lines) to ring up cash registers and adjust inventories. The growth in this field is dramatic, but as this book is written, it remains an area requiring a great deal of experimentation for each application. Those unfamiliar with computer jargon will find it difficult to communicate with knowledgeable engineers in this field. While practical applications are already widespread, one must be certain of the proper equipment matched to the proper job. When used with automatic assembly equipment, it can provide an inspection capability obtainable in no other fashion. (Figure 8-13.)

The machine vision industry has made major strides in producing equipment whose response time allows direct integration.

DATE CODING, SERIALIZATION, AND PRODUCT DOCUMENTATION

Now that verification of product quality is attainable on a 100% basis, at each incremental step of added value using one or more of the tech-

Fig. 8-13. Modern machine vision systems are increasingly user friendly. Such systems can identify features without dimensional characteristics.

niques discussed in this chapter many firms wish to identify products that have been so inspected.

Many firms have chosen to date code only those assemblies passing all functional tests on assembly machines. These date codes or serialization may be stamped, printed, color coded, or laser marked. We are fast approaching a period where in the production of parts subject to regulation, recall, or liability exposure, one may expect to find serialization of individual products coupled with permanent or semipermanent documentation (e.g., magnetic type, floppy disks, or hard printout) identifying the quality tests of that product and the recorded data of these tests. All of this can be incorporated into full MIS systems.

SUMMARY

Properly designed inspection stations can give great assurance that the consumer of parts manufactured on automatic assembly lines will have no complaints. The data from these inspection devices can be used to aid in improving machine productivity and even aid in pinpointing manufacturing problems.

Automated assembly systems incorporating major testing systems have come to be the most significant means for corporations to meet customer's expectations but also to meet ISO 9000 or QS 9000 standards and guard against warranty, recall and liability costs.

Chapter 9
Meeting Government Regulations

INTRODUCTION

Manufacturing is not an island unto itself; it must function within the constraints of governmental regulation, standards organizations, corporate rules, environmental concerns, and liability exposure, all which reflect to some degree the underlying human concern for society as well as the employees' and customer's well-being. This chapter's emphasis is on how to design productive assembly machines within these constraints.

Proper incorporation of quality assurance stations in the initial assembly system and layout discussed in Chapter 4 will do much to ensure customer satisfaction and safety. It is the best defense against future field service expenses, warranty costs, and liability expenses, as well as a defense of corporate reputation.

What is of more concern in this chapter is the negative impact that neglect of the issues until the last moment (often the builder's acceptance run) will have on operator safety and machine productivity. Far too often conformance to regulations and court decisions is left until the last few days of an assembly system project and is never an initial consideration nor an integral part of the machine design. Physical and nontactile guarding and electrical control systems must in-

clude, from the very first moment of design, safety concepts that allow necessary and legitimate access for machine correction by operators or adjustment by technicians as well as operation without jeopardizing operator safety.

Unusual operating environments such as those in sterile or potentially explosive areas or those that generate toxic or hazardous wastes are also the topic of this chapter.

When designing automatic assembly systems, there is a tendency to assume that what is "automatic" runs without human control, intervention, or participation. This remains, for the most part, the star to which we wish to hitch our wagon. The reality is usually far less. Machines have need of human attendants and operators and technicians. That these workers operate in a safe environment was until recent years a social objective secured in a carrot-stick fashion by Workman's Compensation insurance carriers using experience ratings of individual plants to determine premiums.

Many large companies, notably DuPont and other pioneers in industrial safety, tried to instill a concern for occupational safety, an atmosphere in which safe work habits assisted by sound industrial engineering and good housekeeping practices led to a safe work environment. Common sense was used to produce safe work habits. These corporate goals often suffered in practice at the plant level, and smaller companies often depended on the workers' own desire for well-being rather than company policy to maintain a safe environment. In other plants, unfortunately, safety was never a consideration.

OSHA

With the advent of the Occupational Safety and Health Administration (OSHA) in the United States, emphasis was drastically changed from the passive mode of motivating personal habits of safety and identification by signs, barriers, and colors of hazardous areas to a pro-active emphasis on establishing regulations which would require the addition of guards, sound enclosures, barriers, color schemes, and electrical control features that would physically impede the worker from contact with potentially hazardous machinery. These regulations are intended not only to physically bar the worker from accidental contact but to impede his or her access to dangerous areas when done deliberately no matter what the motive.

The philosophical implications underlying these regulations in a country reputed to be an industrial democracy operating in a free economy are staggering, but not the topic of this book. What is of concern here is that the regulations promulgated in the Federal Register concerning the requirements for machine safety can often be interpreted as being in direct or indirect conflict with design principles for high net production on assembly machinery.

Both builder and user should know that there is a basic distinction in who carries the obligation to conform to OSHA standards. Unlike our European counterparts where conformance to national standards for safety is solely that of the machine builder, *American law requires that the user of industrial machinery, not the builder, is required to see that the manufacturing equipment meets OSHA standards before it is operated.* This disparity between similar laws in Europe and our own did not happen accidentally, but came about from a number of vested interest groups who realized that while implementation of the law could

not be blocked, its thrust and subsequent liability exposure could be diverted.

It must also be noted that American laws were to be enforced on a local basis, that is, state-by-state. Some nonindustrial states have even tried to encourage plant relocation within their boundaries by quietly promising slack enforcement of OSHA standards for those relocating to their states.

Today, OSHA enforcement has been lessened; but even in its earliest days it tended, because of lack of standards, not to get too specifically involved in machinery design, other than noise levels, belt guards, and press controls. Had OSHA enforcement been left solely to federal government regulatory agencies, it would probably have not become the problem that it is in those states with strong unions. It is the user orientation toward their responsibility for conformance that has plagued the special machine builder.

Major corporations, being politically sensitive, quickly established OSHA conformance teams to ensure that industrial equipment purchases met newly imposed obligations. These corporations formed OSHA teams, usually without any responsibility for the restrictions their decisions made on machine productivity, established corporate or divisional policy in accordance with their subjective interpretation of the new regulations. Individual states established their own safety standards, tightening the noose further, particularly in the area of noise and environmental protection.

Assembly systems developing on the leading edge of technological developments do not lend themselves to standards. The physical differences and broadly divergent attitudes of leading system builders toward design philosophy made it difficult to establish effective widely applicable industry standards for safety prior to design and construction of assembly machines. And so, we came to a rather pragmatic approach by corporate safety teams: "You build the machine and then when you are through we will tell you how to guard it." This certainly simplified the purchase phase of the project, but often meant substantial additional time after debugging and prior to shipment for the installation of barrier guarding and sound enclosures.

Having provided designs for excellent visual and manual access to facilitate the clearance of jams and to speed the restoration of the machines to operation, builders now found their machines wrapped in barrier guarding which effectively prevented the operator from getting to the machines. The harpies on the shore had once again plucked the eagle out of the sea. On the production floor, machine operators became totally frustrated by their lack of access, removed guards, taped out safety switches, or threw in the towel and asked for reassignment.

We are finally beginning to see in most plants a realization that if assembly systems are to realize their full potential under OSHA regulations, creative approaches to operator safety regulations must be found. These approaches must be incorporated during the initial design phases of the machine and provide for unhindered access where required by the operator, while providing for complete safety and physically barring the operator from areas that should not be available except for major maintenance.

These approaches to operator safety will focus on electrical controls, mechanical barriers, sound enclosures, and optical and electronic barriers.

It is interesting to note that European companies, purchasing American companies or building new facilities, have caused within the assembly system industry radical changes in the design of guarding and in its incorporation in initial system design activities. This attitude is rapidly changing American corporate attitudes.

Operator Safety

Operator safety guarding on complex assembly systems will not work unless the nature of guarding is such that it will not frustrate the operator thereby preventing them from doing their job: achieving significant net production. When operators must spend countless minutes removing awkward mechanical barriers to clear jams or perform other corrective actions they can be expected to seek ways to overcoming these barriers. Rigid guards held in place with threaded fasteners are often permanently removed on the factory floor. Hinged guards are tied back. Electrically interlocked guards will have their interlock switches wired out or taped down.

Any guard that must be routinely physically removed in the normal operation of the machine, specifically to correct product failure rather than machine damage, is an imposition on the operator and a detriment to production. The production motivated operator will seek to remove any such barrier or frustrate its intent thus leaving the operator exposed to hazard, the operating facility exposed to OSHA penalties, and the machine builder exposed to liability claims.

There are better ways. In order to incorporate them, the designer must have a sympathy for the mentality of motivated operators and never underestimate their ingenuity. There is a great deal of truth hidden in the cliché that "anything can be made fool proof but not idiot proof." The corollary is that motivated operators are neither fools nor idiots.

Electrical Control Design. In Chapter 4 the focus was concentrated on the role of the control system in coordinating machine motions and monitoring product quality as well as the controversial choice of how to handle the information supplied by inspection devices monitoring the developing quality of the product.

The operator's initial point of contact in operating an assembly system is the push button control station. At a theoretical minimum the machine will have a START button, a RUN button, and a STOP button. In practice we will also find a JOG or INCH button. This may be combined with the RUN button using a selector switch. The functional role of the jog button is to permit the equipment to be operated under power when machine sensors have stopped the machine and indicated an abnormal condition, a missing part, a jam, or other assembly mishap. These conditions will allow the machines to be jogged without machine damage. It is expected that the circuit design will not permit the machine to be actuated by the jog button if foreseen machine damage will result. The jog button, therefore, will normally be used by operators or maintenance people for set up and adjustment and by operators to withdraw or otherwise move tooling by power in the process of removing jams or defective parts and in the restoration of the machine to normal operation. This work usually involves operator penetration of hazardous operating areas and hence jog buttons are usually wired to bypass both quality monitoring probes and machine safety guarding.

It is anticipated and accepted by most safety personnel that such conditions frequently exist and that the nature of a jog button, a momentary push button, is to be sure that the machine will be stopped instantly by release of the button. Where such instant stoppage is not controlled by releasing the jog button, rigid barrier guarding should preclude access to all but qualified maintenance workers. For instance some operations on an assembly machine such as resistance welding, hydraulic forming, or leak testing may be done by special units housing their own source of power and their own timing mechanism. Their action may be triggered by the main control system, but once started, they will normally continue to a conclusion even if the jog button is released. These operations are for this reason potentially dangerous, and the operator should be kept from these operations by rigid barriers. Figure 9-1 illustrates one such station.

In reviewing proposed electrical design from a safety standpoint, one should look for several potentially dangerous conditions.

One major problem occurs when a large assembly system has more than one push button station. It is common in large machines to have dual push button stations, one on each side of the machine. The system designer should be completely assured that when a machine is stopped by the operator on one side, it cannot be started by an operator on the other side without some concurrence by the first operator. This usually consists of the pressing of a reset button, before the run button will become operative. The addition of an audible horn and a time delay relay will provide both warning and withdrawal time. (Figure 9-2.) The possibilities of accidental restart are enormous, and the construction, function, layout and location of push button stations should be reviewed most carefully.

One controversial area requirement remains as to whether or not a selector-switch-type push button combining both "inch" and "run" facilities in a common button, as shown in Figure 9-3, is safer than individual INCH and RUN buttons and a

separate selector switch. Corporate decisions should prevail.

Mechanical Barriers. *In general, the use of rigid mechanical barriers should be limited to preventing machine operators from areas of definite hazard, limiting such access to maintenance technicians.* (Figure 9-4.) Rigid guarding should not be used where access is required for normal machine operation. Belts, pulleys, sprockets, chain, and pinch points are all typical of areas that are hazardous during normal operation. Not only should these be guarded in a way as to prevent access, but color schemes are often used to warn that hazards lie beneath the barriers. There is a predominant school of thought that barrier guarding should be painted on the outer surface a different color than that of the machine. This color, such as Focal Orange or Focal Yellow, is often one that signifies hazardous areas. Another

Fig. 9-1. Capacitive guards prevent access to this press operation and indicate that hazardous conditions exist when jogging.

Fig. 9-2. In large machines, which require two operators, visibility may be impaired. The use of reset buttons and audible warnings prevents accidental start up by the operator on the far side.

Fig. 9-3. Inch-run or jog-run selector type buttons combine mode choice and actuation in a single button. However, separation of two actuators and one selector button is considered safer by others.

approach is to paint the guarding the same color as the machine's exterior, but paint the interior of the guards and the hazardous moving elements a high visibility color such as Focal Orange. The thinking behind this approach is that exposure to these attention-getting colors should be extraordinary, rather than routine. The appearance of these colors designating a hazard should alert anyone coming to the machine that a dangerous situation exists, such as an open panel door.

In some instances it may be essential to proper operation of the machine that certain hazardous areas be readily seen. Visual access can be obtained through the use of transparent guarding material such as Lexan. Ordinary Plexiglas is not a sufficient guard. Figure 9-5 illustrates a transparent guard.

Sound Enclosures. OSHA regulations concerning hazardous conditions have taken a strong stand on restraining industrial noise. This legislation as written seems to encourage noise containment, but permits the use of such devices as ear plugs and noise muffling headsets such as are

Fig. 9-4. Rigid barrier guards are fitted to areas where access should be limited to qualified maintenance workers.

worn by ground attendants at airports. In practice, however, enforcement of the latter course is often difficult or impossible. Many workers resist any form of ear plug or ear protection. Some find it uncomfortable. Others feel it degrading or a restriction on their personal liberty. Others find it difficult to communicate. Still others feel it prevents them from hearing slight changes in the noise levels or noise characteristics that indicate abnormal operation. Most users feel that the enforced use of ear plugs in any form is so difficult to control that they find it easier to accept the penalties of lost production caused by noise containment practices. Ironically these same workers may use personal radios continuously at sound levels guaranteed to cause hearing losses.

The present OSHA standards set by the federal government in the United States require that noise levels in industry be lowered to 90 decibels. Some states have set 85 decibels as an upper limit, while other legislative bodies and some corporate goals push for an upper level of 80 decibels. Those unfamiliar with the measurement

scale for noise should understand that the scale is a logarithmic one and that the actual noise difference between 80 and 90 decibels is not a 12% difference, but is almost a doubling of the noise

Fig. 9-5. A Lexan guard is used when access should be limited but visual observation of operation is required.

level. Ninety decibels is a practical level for most assembly machinery. Eighty five decibels is difficult to obtain. Eighty decibels is a level below that found in any office using word processors, electric typewriters, or terminal printers. As a practical noise level for the average industrial manufacturing environment, 80 decibels is absurd.

The measurement of noise level is not an easy one. Early regulations did not specify the basis of measurement. The NMTBA established a widely accepted measurement format shown in Figure 9-6. Later legislation has been more specific as to measurement procedures, frequencies, intermittency, and other factors. The machine designer must determine at the earliest stages whether federal, state, or corporate goals will be the governing limits. If these are less than 90 decibels, the question of the availability of waivers on maximum sound levels must be determined. The second question is whether these are peak levels or intermittent maximum levels. It should also be ascertained at this point not only what is the ambient level of the background noise in the area where the machine is to be placed, but also whether that ambient level may change through the introduction of additional new equipment, and its supporting services and equipment.

Ground rules must be established concerning the noise measurement of the machine. Is it a noise level in isolation or one combined with ambient background levels, since noise is additive? Echoes due to location of walls and ceilings and

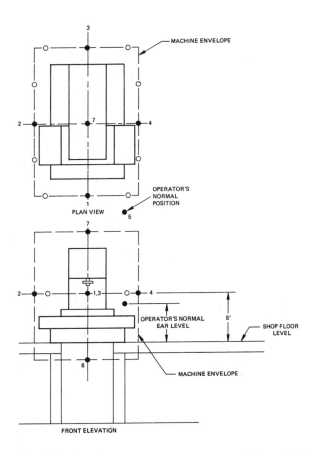

Fig. 9-6. A noise measurement prepared by an NMTBA committee to ensure standard noise measurement practices.

proximity to other equipment may also be considerations. Unfortunately no final readings can occur until the machine system is placed on the production floor.

The problem created by OSHA rules essentially is not on how to contain noise, but rather the containment of noise in a way that permits the machine to be operated efficiently. Figure 9-7 illustrates the problem graphically. The most damaging problem in terms of the downtime that such sound enclosures cause to most assembly machines does not come from failure to insert the part into the work fixture, but from jams within the feeder and horizontal transfer rail. These jams if undetected allow the transfer rail to empty out. Downtime is caused not only by the time required to clear the jam, but by the time required to refill the transfer rail before restarting the machine. A good operator consistently monitors feed rails to ensure that jams are removed while there is a reservoir of parts remaining in the rail. Sound enclosures not only restrict visual access to spot incipient problems at first sign of a jam, but also impede manual access for correction of the problem.

A sensitivity to the problems generated by noise containment devices, such as sound enclosures, should prompt the designer (and customer) to review carefully proposed machine concepts to see if the noise can be eliminated or reduced at the source. Are feeders too large? Are they asked to do too much in orientation efforts that could be achieved more quietly in escapements or transfer devices? Is fluid power used where mechanical action could do the job? Are air jets used to correct poor design or inadequate development?

Are exhaust ports or solenoids piped to a muffler or merely vented to air? Can the use of mechanical feeders or operator-assist units reduce noise generation? Are the new linear feeders capable of efficient feeding and significantly lower noise levels? Are parts feeders equipped with high-low level sensors that turn the feeder off when an adequate supply of parts are contained in the transfer rail? Attention to these details can significantly reduce overall ambient noise levels and reduce the necessity of sound enclosures.

Where noise levels do require sound enclo-

Fig. 9-7. Sound enclosures on large feeders make part loading difficult, and visual access for clearing jams and removal of bad parts impossible for the average operator.

Fig. 9-8. An access port in the sound enclosure facilitates observation and correction without removing the entire sound enclosure.

sures, provision for access ports as shown in Figure 9-8 will be beneficial.

Optical and Electrical Barriers. There are three ways in which an operator can be restrained from entering the operational area of an assembly system in motion. One is the use of mechanical barriers. The second is some form of disruption to electric holding circuits by the physical presence of the operator. The third and most attractive restraint is the use of light curtains.

Physical restraint by the use of mechanical barrier guarding to the functional areas on the machine such as escapements, transfer units, inspection probes, and work-holding fixtures is neither effective nor wise. The frequent need to place hands in these areas to correct or replace defective products or components should dramatically limit the use of rigidly mounted guards. Physical barrier guards that are easily removed but have no electrical interlocking features are worse than useless. If the machine will run with them off, it will be so run. If they are interlocked, only the most ingenious hinges, latches, and counterweights reinforced by strict disciplinary action will keep the interlock switches from being deactivated. Human nature will force the safety conscious designer to look at methods for guarding that require neither time, effort, nor fatigue to stop the machine when the operator wishes to correct some conditions in a potentially hazardous area. The European mode of barrier guarding is finding increasing acceptance in American plants.

Electrically activated sensors which are intended to stop the machine whenever anyone enters a hazardous area are of several types. One approach is to use electrical floor mats that will shut down a machine when anyone stands near enough to the machine to be able to reach into the moving equipment. These floor mats are inexpensive and need only to be placed in series with the holding circuit for the run button. They have not gained broad acceptance for several reasons. They do not really tell if there is an attempt to penetrate the hazardous areas. They make close up visual monitoring of an operating station difficult. Mats can be readily moved, and they often can be easily straddled.

A more effective type of electrically actuated guarding widely used in the 1980's made use of capacitive or radiation type fields generated by a transmitter or antenna. Anything or anyone entering these propagated electrical fields would change the characteristics of the field, and thus stop the machine. Figure 9-9 illustrates such a system. The placement of such antenna must be done empirically after completion of the machine.

Some of the problems inherent in such a system are related to the propagation of the field, and the strength of the field. Since these fields will be essentially circular around the antennas, and proportional to the strength of the field, it is hard to know without a field strength meter whether or not an effective protective field exists in any given area. There are so many moving elements in an assembly machine that the antenna system will have to be so positioned as not to be actuated by normal machine motion. There are points such as push button stations and magazine and magazine feed units where operators are required to have close proximity to the machine. The antenna system has to be detuned in these areas as shown in Figure 9-10.

The absence of a precise demarcation of the barrier line and the possibility of nulls or voids

Fig. 9-9. A—Copper tubing is used for antennas located in the areas where access will be required to produce an electrical field that can detect human intrusion. B—The power supply provides the electrical field and detects changes in that field.

in the field have led to many known accidents where operators have been injured when entering a hazardous area erroneously anticipating that the system would shut down the machine. Today no builder willingly uses or suggests such devices.

Light curtains, long popular on the European continent and later approved by British safety agencies, have had increasing popularity in the United States. A light curtain is a barrier wall established along a specific path by infrared beams or a modulated light source. These light sources are chosen to eliminate the possibility of transient light sources, such as reflected sunlight, giving a false indication. This wall can be established by a transmitting source and an active photo-electric receiver as shown in Figure 9-11 or by a transponder unit using a passive reflector to return a modulated light beam as shown in Figure 9-12. Rigid barrier guarding may be placed above and beneath the light curtain's path, and curtain height is available in different dimensions.

The advantage of light curtains is obvious. The exact location of the barrier curtain is known. They can be placed in close proximity to moving elements of the machine. Visual access to observe machine behavior is excellent. There are no time constraints in entering an operating area on the machine and there is nothing in their mode of operation to cause operator fatigue or cause the operator to want to bypass their operation.

On the negative side, light curtains require a

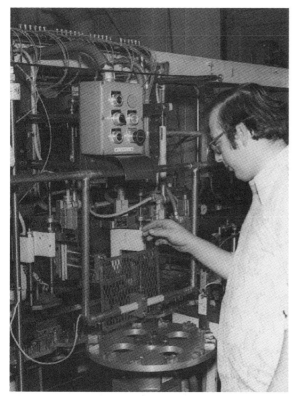

Fig. 9-10. An expanded metal shield allows an operator to actuate the push button station without tripping the guarding system.

clear contiguous path from transmitter to receiver on those systems using an active receiver. Those systems using a passive reflector to return a modulated light source can tolerate an interruption where there is a physical obstruction if an auxiliary reflector is used, but there is no safety protection beyond the reflector. *This requirement for a clear path for transmission of the light beam means that effective use of light curtains will have to be taken into account during initial machine layout and design. It is rare that one can install them as effectively after the machine is complete.*

Toxic Material and Wastes

In an age socially and politically sensitive to environmental pollution, it is still possible to overlook assembly machines as a source of contamination. For the most part that is true. There are two areas of possible contamination that should

be examined: oil emission and the creation of toxic or otherwise harmful gases.

Oil Emissions. Oil contamination on an assembly machine can come from two sources common to assembly machines. Exhaust ports of solenoids controlling lubricated compressed air are often directly vented to the atmosphere. Not only is this a source of noise but also contamination of the atmosphere. Connecting these ports to a muffler system will not only reduce exhaust noise but provide a means of collecting vented oil fumes (Figure 9-13). Spray coating of lubricants and silicones to facilitate assembly or to provide lubrication to a sealed product is a second source of oil-emission pollution. The use of dipping techniques or other curtailment of lubricant spray emission should be part of the design effort.

Gaseous Emissions. Welding and laser drilling and similar activities are two additional sources of atmospheric pollution. Welding can cause atmospheric pollution for several reasons. Materials that are welded in the process of automatic assembly often have coatings to facili-

Fig. 9-11. This light curtain system sends a curtain of light beams to an active receiver. Failure to receive the full transmission because of physical intrusion will stop the machine.

Fig. 9-12. A modulated light source is transmitted in this light curtain system to a reflector mirror and then returned to the transceiver.

Fig. 9-13. A central exhaust muffler in the base of the assembly machine can also be used to collect oil emission from lubricated clean air.

tate welding that are made gaseous during the welding process. Additionally, base metals such as lead also generate toxic gas wastes during welding operations. These must be collected and vented properly. Some welding operations such as TIG and MIG, as shown in Figure 9-14, require the creation of an inert environment for successful welding. Concentration of these gases must be examined for possible toxic conditions.

Laser drilling has gained popularity for its speed and the absence of chips. This does not mean that no waste is created, for the material originally in the area where the hole has been created becomes gaseous and should be removed from the atmosphere by suitable vacuum methods.

It is not uncommon to find the use of a nitrogen stream for cleaning purposes in automatic assembly. This usually can be released to the atmosphere without harm. Ionized air has been often used to reduce harmful static accumulation in parts feeders. Any significant accumulation of such ionized air should be examined.

NATIONAL, STATE, AND LOCAL CODES

In addition to the federal regulations established by OSHA and company purchase specifications for special machinery, the builder may find reference to or determine the applicability of other codes. For instance, a purchase order for a machine to be shipped to Chicago, Illinois, may make reference to OSHA, NEMA, NFPA and Underwriters' as well as to state laws and corporate specifications. Conformance with these codes is a very large part of the engineering and construction costs of automatic machinery. At the same time continuous, rapid advances in electronic controls, guards, and test equipment may not fall into any framework of existing codes, particularly when the time required to establish standards usually exceeds the product life cycles of controls and other electronic or computerized equipment.

Any attempt to standardize assembly system controls has been quickly superseded by broad corporate specifications often differing from di-

Fig. 9-14. A TIG welding head is used to join two lead components in a storage battery. Toxic gases are formed and inert gas is used to provide a shelter for welding. Both should be removed by a well-designed exhaust system.

vision to division and plant to plant. The rapid emergence of each generation of programmable controllers and the significant differences in modular construction of new electronic controllers further weakens corporate electrical specifications.

NEMA & NFPA

NEMA standards are industry standards for the construction and capabilities of various electrical devices such as motors, controls, and enclosures that hopefully qualify them for operation in certain applications or environments. These standards are a joint effort of the National Electrical Manufacturers' Associations.

Underwriters'

Underwriters' Laboratories is a quality monitoring activity whose intent is to certify that electrical devices have been examined on a continuing basis and found to have standards of construction that ensure they are not hazardous in their intended use.

With the exception of the area of Chicago, Illinois, there are few reasons to look for Underwriters' approval on industrial controls. Where it is necessary to obtain Underwriters' approval of a specific machine control system, it will be necessary to obtain this approval by obtaining a facilities approval, type approval, and specific approval for a given panel. This is a long and costly procedure.

Local Codes

There are many state and local codes that may bear on machine construction, but these are usually observed quite loosely. The notable exceptions are the requirement to have Underwriters' approval in Chicago, specific noise levels lower than federal standards in some states such as New York, and the highly restrictive Los Angeles County code. It should also be noted that the Canadian CSA standards for electrical construction are quite demanding.

All of these codes are intended to improve safety. Some look to standards of machine construction, others look to product construction quality in the electrical industry. Others concern themselves in reducing liability exposure. Others

are subservient to local social and political pressures. For the machine builder they pose a pitfall if specified in the purchase order and accepted without recognition of their full implications.

CLEAN ROOM SPECIFICATIONS

Occasionally, a builder is requested to quote on a machine to run under "clean room" conditions. A determination must be made as to which type of clean room environment is indicated. Is it a question of bacteriological contamination, particulate contamination, or petroleum distillate contamination? The best source of the actual specification involved is the customer's own quality assurance department. It has been a continuous problem for the clean room industry to obtain a commonly used terminology for clean room specifications.

For the builder, clean room operation can present certain design problems. One often raised is the ability to disassemble and wash those machine areas that contact or are adjacent to the product. Extensive use of stainless steel and plastics is helpful.

Another area of concern is contamination of the machine and the product by the assembly machine's lubrication system. Increasing use of antifriction ball slides, coating materials with natural lubricity, and extensive use of shields are helpful but add significant cost. Most experienced assembly system builders offer lubrication-free equipment, usually at identifiable cost.

The third area is that of escaping compressed air, particularly lubricated air. The use of air jets and exhaust ports vented to the atmosphere should be avoided to the fullest extent possible.

A very extensive system debugging activity must be accomplished prior to final installation because of the difficulty of further modification and extensive system repair or rebuilding in the clean room during early production runs.

EXPLOSION-PROOF OPERATION

A significant number of assembly machines are involved with a concern for explosion-proof characteristics. Machines assembling blasting caps, electrical detonators, thermally ignited devices, and military items such as fuses and grenades require specific treatment. One way to provide explosion-proof capability is the necessity of eliminating wherever possible the use of wiring external to the main control panel. For instance, air-operated vibratory feeders solve one major problem. But it is not enough to assume that design attempts to prevent explosions are sufficient. Machines must be designed with the possibility of accidental product detonation in mind. When explosions do occur, they should not be confined but diverted in a direction that is harmless. Every possible effort must be made to prevent countermining or sympathetic detonation in other areas of the system. There is only one way to be certain of safety in design and that is to actually do controlled detonation experiments. The cost of the machine should include possible station mockups for detonation experimentation.

Static Dissipation

One ever present problem in the assembly of products which can be accidentally ignited is the accumulation of static charges. The concern over static dissipation, however, extends to other areas. Assemblies with light plastic and rubber parts are often hindered by static accumulation. Many electronic components such as LSI devices can be severely damaged by static charges.

The first way to avoid this problem is to avoid the buildup of static charges. A determined effort must be made to secure grounding of all machine elements and electrical bonding of the machine to a grounded floor. The use of plastic and rubber in conveyors, transfer tubes, and magazines should be avoided. Ionization of part pieces may prove useful.

Intrinsically Safe Relays

It is difficult to design a machine without switches but the limited number of available models of explosion-proof switches is frustrating to the machine designer. The size of explosion-proof switches, their weight, and lack of sensitivity are inhibiting.

Fortunately, the advent of intrinsically safe relays now permits the use of normal push-button switches. The current level on the external conductors is in the range of milliamps and, in most cases, insufficient to cause accidental ignition. Figure 9-15 shows an intrinsically safe relay mounted in a control panel.

There is a reluctance on the part of those acclimated to working in explosive atmospheres or environments to deviate from electrical devices specifically manufactured for explosion-proof use, but the development of the intrinsically safe relay and its approval for specific levels (Classes E, F and G) of explosive environments add new versatility to assembly machine design and construction.

Positive Air Pressure

One safety feature often incorporated in machines working in areas where accidental detonation is a hazard is that of placing a light positive air pressure in control enclosures. This prevents explosive dust from entering cabinets and enclosures with faulty seals.

The designer who is new to such construction must be made aware of the incredible pressure exerted in such cabinets by very light pressures. One pound per square inch will exert over one ton of pressure on a door that measures three feet by five feet.

THE IMPACT OF PRODUCT LIABILITY

Probably no single judicial or legislative action in recent years has had so much impact on the machine tool industry as that of the concept of strict liability. Until recent years, product liability was effectively limited to contractual parties, that is, the buyer and the seller. For the most part, lawsuits pertaining to machine tool performance were rare.

Today anyone in "the stream of commerce" can sue or be sued for damages—fiscal, physical, and psychological, real and imagined. An employee using his or her employer's machinery was once thought to be limited to workmen's compensation if injured on the job. Today the courts accept suit, when the machine is not only a direct cause but even a proximate cause of injury, filed by an injured employee. He or she may direct his or her suit against his or her employer, fellow employees involved in the purchase of the machine,

Fig. 9-15. Intrinsically safe relays allow use of normal commercial switches in areas which formerly required explosion-proof switches and potted conductors.

dealers, agents, and manufacturer's representatives involved in the sale of machinery, the assembly system integrator, component vendors, and employees of the machine builder.

Assembly System Builders

In recent years builder's premiums for liability insurance have risen by as much as 2000% or 3000% and some machine firms, particularly press manufacturers, find themselves unable to obtain coverage at any price.

There are at the present time no effective statutes of limitation, and liability exposure remains, no matter how old the system or machine, no matter how many prior owners, and no matter how well it is maintained or used.

Machine tool builders are always hopeful that new legislation will provide more equitable guidelines for resolution of liability claims. In the meantime, however, machine builders remain concerned about the performance and safety of the equipment in the field. Much of the concern

of assembly machine builders in discussions over guarding concepts lies in their desire to avoid potential liability suits by providing guards that the operators can live with, guards that will remain in place and remain operative through the life of the machine.

The Machine User

The assembly system builder is not the only one with product liability problems. Assembly system users face enormous problems with the products they produce not only of field service and warranty costs but with government-imposed recalls, class action suits, and specific liability lawsuits.

There are only two useful defenses against such product liability lawsuits. They are documentation of quality at time of manufacture and proof that all the actions a prudent person could take in the manufacture of the product were in fact taken.

Documentation of Quality. The assembly system's unique ability to inexpensively monitor each and every product assembled has been mentioned throughout the book. The very nature of part feeding is in itself a vital product quality verification. Monitoring of product function after assembly is another defensive weapon. Assembly on a machine equipped with well-thought-out inspection units that verify component presence, component position, and functionality of the assembled product is a strong defense against product liability lawsuits.

The "Prudent" Rule. The courts expect that when offering a product for sale, the manufacturer exercises every precaution a "prudent" person could take in producing that product. Documentation of the design effort that went into the machine inspection system can be of great assistance as demonstration of prudence.

SUMMARY

The advent of OSHA caused great concern in industry. Today one hardly hears about OSHA, taking it as a condition of life. Political pressures and economic realities have slowed down new environmental legislation. The question of product liability remains a major problem for both system builder and user. Safety in all its aspects must be built in the system design from the very first concept phases of system development.

Chapter 10
Systems Procurement

INTRODUCTION

The purchase of an automatic assembly system requires major decisions on the part of the purchaser. The goal of such procurement can be spelled out simply—to assemble the largest quantity of the highest quality product at the lowest cost. Time-to-market, short product life cycles, inability to forecast volumes, and market changes, however, preclude leisurely development of assembly machinery, especially when delays in obtaining capital funding for identified opportunities occur. Current management attitudes toward product quality and time-to-market have somewhat eased delays in capital approval.

Often the exhausting work of securing management approval and funding for the purchase of an assembly system leads to a psychological letdown of the procurement team when final approval to release a purchase order is given. This quasi-lethargy often means that the most critical times in assembly system evolution, namely purchase order issuance and acknowledgment, concept development, machine and station layout review, often pass by without enthusiastic and full participation of the customer purchasing group.

Establishing an equitable contract, particularly when doing concurrent engineering for a one-of-a-kind system, requires a mature examination of the intent as well as the letter of the contractual agreement. The purchaser must recognize that by his or her selection of a given vendor he or she establishes unique and subtle ties to the chosen vendor. *Success will require a skillful combination of both cooperation and isolation* of the purchaser and the vendor.

Failure to establish a full determination of intent on both sides often lies at the heart of unsuccessful ventures into automatic assembly, particularly when the purchaser often feels a technical inadequacy to interact with the machine and control system designers during the concept stage.

While the builder team hopefully knows more about automatic feeding techniques, automated testing, data acquisition and integrated systems, there is no reason to suppose that they are fully knowledgeable of the customer's product field. As an engineer or, more often, a tool designer, the vendor may be somewhat insensitive to probable customer product quality level requirements, normal part variations, possible future product changes, and operating considerations. *The success of an assembly machine project will result from both sides contributing from their fields of competence.*

PLACING THE PURCHASE ORDER

Until the advent of simultaneous engineering, the purchase order for an automatic assembly machine was intended to be a brief, legally correct statement of the result of a series of actions including project definition, requests for quotation, reviews of proposal, and a multitude of discussions capped by a successful quest for capital funds. It is a legal document which should attempt to define the intent, obligations, and responsibilities of both parties. In truth, purchase orders rarely meet these standards. In Chapter 1 reference was made to Powell Niland's work on the acquisition of special machinery. It is worthwhile repeating several comments in the abstract of that book: "One of the most important conclusions emerging is that there are substantial benefits to be realized from careful preaward planning, *especially by working up soundly conceived specifications in considerable detail. Some of the more frequently encountered difficulties as well as some of the more serious, had their origins in failure to perform adequate preaward planning.* Efficient communication between representatives of the buyer and vendor also emerged as a critical need. Finally, the author concludes that in acquiring this type of equipment there is some advantage in restricting purchases to a relatively few, carefully selected vendors with whom the user can work in close cooperation at all times during the acquisition process." These specifications must include not only the assembly tasks, but required testing systems and MIS requirements. These specifications must be functionally defined. The impact of simultaneous (concurrent) engineering on the structure of the purchase agreement deserves much study.

Defining the Objectives

The objective of any capital expenditure should be to increase operating profits. The ultimate related goal should be to achieve the maximum profitability by the minimum combination of capital expenditure and operating costs. To achieve this takes a measure of objectivity, knowledge, and future vision not often found in one place. One way to achieve these objectives is to formulate in writing the specific targets of any assembly improvement project in very definite terms. Cost reduction, improved capacity, improved delivery, and improved quality are some of the obvious targets.

A concise statement of present problems and possible goals can be supplemented usefully by a concise statement of the probable levels of support for the attainment of such goals in every area of management. It is rare to find enthusiastic uniform support for new production techniques throughout management. It is equally useless to propose solutions that cannot work in specific manufacturing environments.

Management by objectives is a recognized technique, but it assumes the objectives are attainable. The problems of poor assembly productivity may be technical, they may be financial, but the problems can often be a failure to obtain a sufficient management consensus as to the actuality of the problem, the feasibility of its solution, and the determination of the whole management team to take the necessary steps and enforce the new disciplines essential to bringing an assembly mechanization project to fruition. *Countless studies of automation efforts point out that the greatest implementation problems are managerial, not technical.*

Proper Use of Corporate Specifications

Most large firms with formal purchasing, industrial engineering, and manufacturing departments have prepared some type of specifications covering machine tool procurement. These specifications, like many governmental laws and regulations, start out pure in intent and were usually created because of past abuses. Standardization for maintenance parts is a legitimate concern. Definition of specific design practices for machine construction is intended to reduce maintenance and improve reliability. Specification of work area dimensions to ensure operator accessibility and safety is certainly rational.

The same positive comments could have been made when the graduated income tax was first enacted into law. Specifications, however, continue to grow in scope and complexity. Whole departments, *usually without any bottom line accountability*, have been created in many large companies to improve and promulgate such specifications.

The established system builder is usually uncomfortable with such all-encompassing specifications, and this is for many reasons. Proposal preparation for special machinery is expensive and time consuming. Not the least of the proposal costs is the time spent after the builder's technical solution and project costs are determined in reviewing the proposal in light of formal customer specification to determine additional costs. These costs may do nothing to improve machine capability and, in fact, may cause areas of the machine to deviate from proven successful practices in automatic assembly.

Corporate specifications are often in conflict with attempts to create nationally recognized industry standards. The former JIC electrical specifications are a typical example. These standards were established at great industry cost to ensure uniformity of electrical construction practices in the machine tool industry by a group of electrical manufacturers, machine builders, and machine users. No sooner were they promulgated than new corporate specifications were issued to complement, expand, and change the JIC specifications. In some major companies, different standards for electrical construction practices exist at corporate staff level, at divisional level, and at different plant sites within the same division, with the differences including brand preference, design practice, and safety considerations.

Very often, corporate or divisional specifications are based on prior relationships with specific vendors. If these relationships are objectively based on past experience concerning reliability, field service, and parts availability, they are possibly valid. If the vendor/customer relationship is based on social considerations, they are a poor basis for major design decisions.

Not only do corporate specifications impinge on nationally established standards, but they also assume that builders have not established standard industry practices within their own firms, standards established on past experience, field service reports, and manufacturing efficiency. If a builder's standards are shoddy, that is a sound reason for not selecting the builder as a potential vendor.

Deviation from an experienced builder's normal practice imposed by rigid customer specifications generally will increase the machine price without necessarily improving machine efficiency and reliability.

Is there no validity to corporate standards? If we assume that specifications reflect current building practices and modern material and technologies, and if they do reflect attempts to prevent chronic manufacturing problems, both operational and maintenance, they can be used as realistic check lists for design review purposes. Specifications used as minimum acceptable levels of machine quality are very worthwhile. But to the extent that they preclude the vendor's ability to offer innovative design and force the builder to deviate from established proven practices they remove the burden of operational reliability from the builder and place it on the customer. Most importantly, the time spent in most specification development processes usually means specifications are obsolescent by the time they are released and promulgated.

Seeking Quotations

The quality of proposals from machine builders is often matched with the quality of the proposal request.

Preparing the Request for Quotation. Since assembly machine builders are, by the very nature of assembly system development, risk-sharers, they are naturally interested in reducing the unknowns in any assembly project. Adequate information in the request for quotation will reduce the time required for quotation preparation and will usually enable the builder to reduce his or her quoted price. The request for proposal should include as many of the following items as possible:

- Sample components and/or part prints.
- Annual and seasonal volume requirements and estimated length of model run.
- Possible variations or combinations of the main assembly, or possible future design changes.
- Comparison of the project requirements to other assembly work in the organization.

- Nature of required inspections on the partially completed and completed assembly.
- Number of component sources.
- Degree of freedom to modify design of components to facilitate assembly.
- Extent of prior experience in automatic feeding of any component.
- Target date for preproduction runs.
- Target date for production runs.
- Surface-finish requirements that might affect feeding and fastening.
- Prior and subsequent operations.
- Environmental operating concerns.
- Availability of maintenance personnel.

Selecting Prospective Vendors. It is neither fair nor ethical to request formal proposals on complex assembly systems from a large number of builders. The cost of preparing such quotations is extremely high. Firms that regularly request large numbers of quotations without significant purchases will find the more reputable builders either reluctant to quote or unwilling to devote detailed attention to their requests. In the long run, such firms will find that they are spending a lot of time in trying to evaluate large numbers of quotations. They also run the risk that the more successful builders will be unwilling to take the time to close-price their quotations. It is far better to spend both time and expense in evaluating and, if possible, visiting potential builders before releasing requests for quotation. Three or, at the most, four builders should be selected to quote on any specific job.

Criteria for selecting these systems builders would have to include past experience in the building of assembly equipment, and, more specifically, experience in similar or related types of assembly. Historically, established systems builders have tended to concentrate on particular industries or specific types of assembly.

The financial condition of the builder is critically important, particularly in concurrent engineering programs. Some customers have found themselves forced to financially bail out marginal builders to obtain completion of needed machinery. Work load or order backlog in a builder's plant may also be a deciding factor. Reputation in maintaining delivery schedules is important;

however, some latitude is reasonable here. The research and development requirements for assembly systems can often lead to unforeseen delays, particularly in a simultaneous engineering environment.

Evaluating the Quotations. A broad spread in quoted prices is common to automatic assembly system quotations. The customer must carefully determine what portion of this variation is due to the builder's internal efficiency and skill, and what is due to actual differences in the design and capabilities of machines offered. One builder may feel a more expensive control system will increase net production. Another may feel additional automatic gauging should be incorporated, increasing machine costs but reducing later inspection costs. *There is wide variance in post-delivery support included in the quoted prices by system builders.*

The first step in evaluating the proposal is to compare the quoted prices and determine exactly what is offered for that price. But this is not enough. A second factor is probable net production. A third is estimated time between delivery and release of the proposed machine to full production. Some experienced buyers of automatic assembly machines have gone so far as to set up expected lead time periods for each of their builders.

The customer is ultimately concerned with assembly cost per unit of production. It may be less expensive to pay more for a fully automated machine and obtain greater efficiency. On the other hand, it may be less expensive and more productive to retain certain hand operations, ergonomically coupled to the automated system.

It is not uncommon on large complicated assembly projects to ask for modified quotations after initial evaluation. It is desirable at this point to make sure that both sides have an absolutely clear picture of what is being asked and what is being offered. Ford's traditional PIR/PN system worked well. A request for quotation designated as a purchase information request (PIR) asked builders to evaluate the technical feasibility, probable cost, production rate and delivery of a project to determine if Ford management would approve the project. A purchase notification (PN) request meant that it was a funded and approved program. PN specifications may include or combine many of the specific unique features described in various PIR responses to get the best possible combination of features quoted in each vendor response to the Ford purchase notification.

Determining Net Production. What's the probable production rate? As assembly machines become more complex, this question becomes harder to answer. For a first machine, the builder can often offer no more than an educated guess. The user, on the other hand, desires some definite minimal rate of production on which to base machine justification. This area will continue to be the most sensitive one in the procurement of assembly machines.

Certain guidelines can be established. Production obtained during the builder's trials should not be considered the final rate. A gradual increase in realized net production can be expected for some months as operators and maintenance people become more familiar with the equipment and minor improvements are incorporated.

An increasingly strong tendency among reputable (i.e., successful with a backlog) builders is to guarantee only gross-production capability. This condition is coupled to insistence on a first acceptance run at the builder's plant. The corresponding quotation will read something like this: "This machine will produce at a rate of X assemblies per minute." This statement means that the machine, with its associated transfer devices, feeders, and inspection and control systems, is mechanically and physically adequate to *continuously operate* at the gross cyclic rates quoted. *Nothing else should be read into this*.

The builder can be only held *totally responsible* for station and system reliability. The builder is not able to control either the quality or consistency of component parts coming to the machine, nor the training and motivation of user's personnel. By insisting on first acceptance runs on the builder's floor, the builder is able to control these two outside variable elements sufficiently to demonstrate fundamental machine capability. This first acceptance run should not be meant to

relieve the builder of installation responsibilities or initiate machine warranty terms.

Establishing Payment Terms. Automatic assembly systems can be very expensive. Deliveries are extended and often substantial amounts for purchased items such as lasers, vision systems, etc., are spent early in the machine development period. This trend is increasing with the greater use of computers and the need to develop both operational and integration software.

These two elements create working capital problems for builders. Except for a very few builders who are divisions of major corporations, most firms in the automatic assembly field are of modest size. For this reason, many builders often request some form of partial payment or progress payments when quoting on large machines. Ever larger projects and simultaneous engineering projects make this critical. In the event such progress or milestone payment terms are not authorized and the builder is forced to borrow, it is inevitable that the cost of borrowing will appear in the price of the system.

One form of progress-payment term that is accepted by government agencies is to have the builder bill up to 70% of the actual costs incurred during each month of construction, with the remainder due on acceptance. A further provision is that the sum of these progress payments shall not exceed in total 70% of the total established purchase price of the machine. All invoiced costs must be subject to outside audit at the buyer's discretion.

This is only one such form these terms can take, but this particular form provides the buyer with a check on the builder's progress. Other terms may be based on milestone achievements. For instance, terms may read:

- 25% due on completion of engineering
- 25% due on completion of fabrication of tooling
- 30% due on approval for shipment
- Balance due on final acceptance

Many special machine customers are reluctant to pay progress on milestone payments and insist on terms of net 30 days, or a specific percentage when the machine is approved for shipment and balance due when final acceptance is made on the floor. The increasing number of European and Asian owned plants in the United States that are used to progress payments may cause American companies to modify their rigid payment terms. The scope of current programs pushing machine builders into total systems integration in a simultaneous engineering mode will demand modification of this rigid stance.

Payment terms agreed to at time of order reveal a great deal about the vendor and the customer. There are many subtle implications in payment terms and most quoted terms are negotiable. The customer should be wary of any suggested payment terms when cash advanced precedes or exceeds work done on the project. Cash payment in advance of work performed usually leads to strained relationships. It may foreshadow the builder's inability to pursue the project through to completion or to provide the post-installation service which is essential to full machine development. Even the strongest builders may lack adequate capital for very large systems in a simultaneous engineering project without some form of progress payment.

Customers must realize, particularly in times of high interest rates, that builders may expend the majority of the overall program costs several months before delivery of the assembly system. Without progress payments, even the largest system integrators may defer timely deliveries of essential purchased parts to the detriment of orderly machine development.

Terms may indicate the vendor's opinion of the customer's ethics in meeting obligations promptly. While "net 30" terms seem to offer the greatest security to the customer, terms of 80% or 90% on acceptance for shipment and the balance on final acceptance offer significant security and control of the project to the customer and equitable reimbursement to the builder.

Whenever milestone payments or progress payments are made, the customer's project engineer authorizing payment in effect gives at least tacit approval to the builder, and begins to assume responsibility for performance and delivery. If the builder is late, incompetent, or uncooperative, pressures begin to mount on the project engineer to accept the machine before it is suitable.

Completing the Evaluation. Price, production, and delivery are the essential points in evaluating any possible procurement. Also, when automatic assembly procurement is being considered, the buyer must include in the purchase cost the cost of liaison trips and sufficient sample parts for feeder development and machine tryout. Whenever possible, the user's operating personnel—those who will be directly concerned with the day-to-day operation of the machine—should be present in the builder's plant during final debugging. This somewhat costly procedure should greatly reduce the time lapse between delivery of the machine and final release to production.

The design and fabrication of the automatic assembly equipment is only one side of the equation of incorporating automatic assembly into the production process. Whether this equipment is to be built in-house by the user, engineered by a consulting firm, or purchased as a completed machine, no amount of design skill is sufficient in itself to ensure profitable automatic assembly. The selected system must be capable of operating in the user's environment. This does not relieve machine builders from building to the highest standards of design and construction.

Buyers have the responsibility for selecting builders capable of meeting these standards and for providing them with all available information and aid. The outside builder should be willing and ready to share the risk of special machine development, but cannot and should not be expected to share risk in those sectors of the buyer's operation over which the builder has no control.

Writing the Purchase Order

The written purchase order for an assembly system should be the culmination of countless discussions and vendor selection meetings. It should contain several elements, among which should be a comprehensive statement of the work to be done, operation by operation, and the specific cost of each quoted function. In addition, the purchase order should spell out the terms of engineering concurrence, machine acceptance payment, installation, documentation, training and post-installation service and warranty.

Preaward Conferences. Many companies have adopted a preaward conference in which photocopies of the proposed purchase order are edited line-by-line by representatives of the user's purchasing, industrial engineering, quality controls and MIS group, in addition to manufacturing engineering departments together with the selected vendor. At this time the builder must be required to identify an initial specific technical solution for each proposed part feeding, inspection, and joining operation, as well as to be able to isolate and list the cost for each machine function. This breakdown of costs is essential to controlling any price changes which often occur during the course of automatic assembly machine projects. It is not uncommon to see that changing market requirements, new governmental regulations, or changes in product design and fabrication require modification to the assembly machine system with resulting purchase order changes, deletions, or additions to proposed stations.

If the machine price is given as a lump sum, necessary price modifications during the development of the machine can lead to abuse or, equally as bad, the suspicion of abuse. Since any assembly project is a risk-sharing venture, mutual trust is essential. Mandatory price changes that cannot be adequately explained destroy this trust.

It is also imperative that realistic schedules for specific milestones be established, component parts availability for machine development be verified, and lastly, *that gateskeeper spokesmen for builder and customer be designated for necessary communication during the project.* There is a tendency for project engineers to talk to the machine designer, purchasing agent to talk to vendor sales engineers, plant electricians to talk to control designers, and none of these keeping other parts of the team informed. Failure to insist that communication be passed through a designated spokesman or "gateskeeper" usually leads to delay and aggravation.

During any preaward conference there will be two key topics that will cause a great deal of discussion: delivery and acceptance conditions.

Delivery. Machine delivery is critical to many projects. Late delivery means lost profits, unanticipated costs for interim tooling and project delays. It is worthwhile to consider the question of delivery in detail. It has long been an axiom of the machine tool industry that sour tempers caused by late deliveries are quickly sweetened by the increased productivity of a successfully delivered and operating machine. This axiom, however, is often put to a severe test as late deliveries which have become increasingly common-place are further compounded by prolonged startups, poor initial productivity, and failure to meet projected ROI figures. The encouraging fact is that these problems need not happen. Delivery delays can be minimized or even eliminated through advance planning.

A check list is offered here to help form a delivery schedule for a medium-complexity automatic assembly machine in the $300,000–$1,000,000 range. This guide in chart form (Tables 10-1 and 10-2) is based on a typical sequence of events in the building of a newly developed automatic assembly machine. Duplicate machinery (with no modifications) can be done far more rapidly. In addition, the guide also assumes normal lead times for purchased components as well as builder competence and transferable experience with machines of similar requirements. The chart is to be read upward, starting from the bottom. It emphasizes a sequential approach to system completion and installation and shows a series of milestones that, if not met on projected schedule, indicate that the delivery time is in serious jeopardy. One of the chart's limitations is that it does not provide for nor anticipate significant product or component change except in the earliest stages.

Only one aspect of a dual problem is treated here because, to put things in proper perspective, the subject of on-time machine deliveries and meeting production requirements after delivery will be considered as separate, although related, issues. The first, how to bring the machine online on schedule, will be treated in this chapter. The subject of how to ensure rapid development to maximum productivity after installation is to be discussed in detail in another chapter.

The very term "delivery date" is interpreted differently by builder and user. *Builders often consider delivery as the day they are told to ship; customers usually consider delivery as the day the machine is delivered and completely erected on their floor. The difference in dates may be as great as one month or more.*

Pre-planning can be used effectively only if both the customer and builder have the technical capability and both fiscal and human resource projected capacity to meet the schedule. Capacity is not only a builder limitation. Delivery times at customer sites are very often determined by new model introduction dates. Intervention of regulatory agencies on the national and state level, particularly in products involving fuel economy, pollution or consumer safety, additionally may require submission of sample production parts for certification long before the user can begin shipment to the marketplace, often in direct conflict with machine construction schedules.

Pipeline or in-transit delays may further reduce the available builder's time for machine completion. Many users of automatic assembly equipment do not ship their product to the ultimate customer, but instead must ship their product to final assembly plants. For example, fuel injectors may be made in a plant remote from engine assembly which in turn may be in a different site from final car assembly.

The vast majority of assembly machines produced in this country are for new product applications rather than for cost reduction on existing products. Production sample parts are rarely available for machine development in any leisurely fashion. Product and component redesign during machine construction has become a way of life, particularly in those organizations attempting simultaneous engineering.

Many delays associated with the machine construction schedule can be traced to the customer. For example, isolation of the product design and manufacturing engineering operations at the customer's plant will often disrupt the proposed delivery schedule. The customer must be able to provide product design, industrial engineering, and necessary operational specifications in a time frame consistent with the delivery schedule, or late delivery is certain. If there is any question as to the ability of the customer to provide infor-

mation on time and provide samples when required for machine development, the customer should provide for possible interim manual assembly.

Many so-called "late deliveries" never could have been delivered on the first specified date because of a lack of project definition. The customer's manufacturing engineer assigned to an assembly machine project should not only monitor the builder's conformance to any agreed on schedule but must use the schedule shown in Tables 10-1 and 10-2 to ascertain that their own in-house product definition, component part availability, and site and personnel training preparations are in phase with the building program.

Vendor backlog is often the key to realistic delivery promises. Obviously, essential to on-time delivery is the builder's technical, physical, and financial capacity to bring a job to completion at a specified date. While there are many exceptions, a busy established builder will be more likely to meet the schedule than a small builder or new entry into the assembly machine field. Established builders with significant backlogs can avoid troublesome or difficult jobs by slanting their proposals to offer attractive delivery to those jobs best suited for automatic assembly and their own experience base. Firms trying to enter the assembly automation field are forced to take more difficult jobs or accept unrealistic delivery schedules in order to establish a backlog. The best delivery safeguard for the user is to take a close look at the builder's backlog.

A builder's backlog divided by the builder's average monthly shipment level is the best indication of probable delivery dates. But beware; formal backlog figures do not always indicate actual backlog. Because of possible delivery problems, the tendency of many experienced users is to place orders for production equipment before the product and production process requirements are fully established. This often means that the scope of the project may significantly increase as the work advances to the delivery date. Builders can quote delivery only on the basis of existing orders. The author's own experience is that *most large orders grow in cost (and complexity) on new product projects by 20–30% during the build phase of the assembly system.*

Debugging an automatic assembly system is a fact of life and always is the great unknown. It is the time for proof of the concept, design, and execution of the machine. Much of the criticism of low machine productivity and extensive startup time is directly attributable to a lack of patience during the builder's debugging phase.

Management pressure within the customer's organization (due to business plan and budget pressures) to have a machine delivered on time, without regard to its operational status, inevitably leads to extended delays or even complete failure to realize ultimate production capability. The debugging time periods shown in Tables 10-1 and 10-2 are averages for machine systems of medium complexity. They must be adjusted for project complexity.

Deliveries agreed on in all good faith by both sides when backlogs are extended beyond one year can become increasingly unrealistic as other outstanding orders of the builder continue to grow in magnitude without proportional delivery relief. Failure to recognize the strains imposed by the rapid business expansion shown in Fig. 10-1 leads in turn to failure to realize how easily deliveries can slide without continued monitoring and participation by the customer from order to delivery.

Machine builders have as much need to ship and invoice as any assembly system user. The builder's natural tendency is to favor and push those projects that are clearly defined, to complete those machines where customer product samples are available on time, where emerging questions are answered promptly, and a continued sense of customer urgency and participation is evident. Thus, the *customer that exhibits urgency without providing project definition and guidance may be given lower priority.*

No industry can expect to adjust to the fluctuating market conditions illustrated in Figure 10-1 without some period of adjustment. In a de-

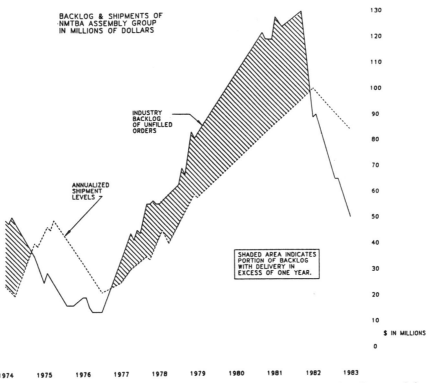

Fig. 10-1. Shipments and backlog levels of the Assembly Machine Group of the NMTBA from 1974 to 1983. This cyclical order/backlog situation is typical of the capital equipment marketplace.

pressionary period, limited market capacity, loss of skilled help, inflation, and part shortages inevitably delay the return to normalcy in the machine tool industry during the early phases of economic recovery.

To compound the builder's problem of market fluctuation, customers facing changing market requirements also have crises within their own companies. Continuous reassignment of customer's liaison personnel, often occurring midstream in the project, interrupts the continuity of experience. Key customer personnel at the end of the project may be different from those at the beginning, and thus may have problems grasping or agreeing with the overall concept agreed to at time of order.

Readjustment of priorities, schedules, and rapid redesign of product requirements by the user further plague both the builder and customer managements. These problems are most serious and threaten continued orderly development of the automatic assembly machine industry.

Late delivery is a dramatic industrial weakness that can be overcome and eliminated by cooperation between builder and user, but this is only one-half of the full story. Failure of the delivered machine to realize acceptable net production levels is a problem whose seriousness is never noted until too late.

Failure to face up to delivery problems, backlog induced delay, design time, build time, and debugging time as well as installation and startup delays does not mean the problems will disappear. If they are not faced squarely at time of purchase order, they will cause real problems during the debugging phase of the machine, a time when mature patience is required most.

If delivery is a critical consideration when writing a purchase order, machine acceptance criteria are even more frustrating. The potential user seeks guarantees the builder cannot in good conscience give them. The buyer wishes to be assured that the machine is capable of sustained operation at efficient operating levels. Acceptance criteria must be mutually agreed on if there is to be a valid contract. At this point many builders face a real problem with their own ethical standards. Acceptance runs on the builder's floor are at best artificial. Sample part costs and logistical (availability) prob-

lems mean that it is highly unlikely that most customers can ever supply enough qualified parts, particularly in concurrent engineering programs, to obtain a prolonged builder's run of any real duration. Most component parts available for machine development, particularly new product development, rarely meet all print specifications.

The importance of the builder's past performance and previous experience is a vital element in the purchase order decision.

Acceptance Runs. The builder's acceptance run must cover several areas. The basic machine must run as a machine. The control system must initiate and sequence actions, and inspection stations must detect specific malfunctions. Transfer stations and joining stations must transfer and join. All these are the builder's responsibility and often can be demonstrated with little or no significant consumption of product part pieces. For instance, a stipulation that the machine operated without product components must cycle continuously for a period of 48 or 72 continuous hours without any stoppage is a realistic and sensible check of system durability, the quality of bearings, controls, cylinders, and similar industrial components. Inspection stations can be verified by running through component parts, subassemblies, and assemblies with known quality levels.

These preliminary runs leave until last the ability of the machine to handle the variations and permutations of the component part pieces that, assembled together, form the product.

The question of machine acceptance is described in detail in the next chapter. The scope of the acceptance run should be detailed in the purchase order.

Machine warranty is an integral part of any order, and the scope of the warranty should be explicit.

The Worth of Progress Reports

Many purchase orders include a request for progress reports. Unless one is willing to establish a critical path chart, most conventionally used reporting forms are inadequate to graphically show the actual progress. The problem is compounded by the sheer inability to adequately and accurately predict debugging for any given

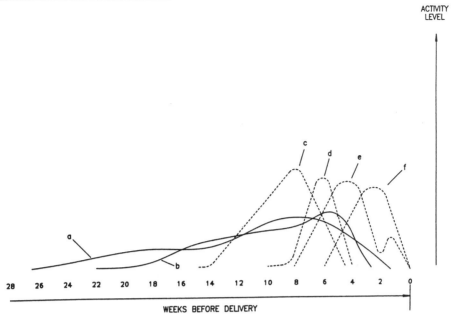

Fig. 10-2. Labor intensity curves of various phases of assembly machine construction: (a) engineering hours, (b) tooling purchases, (c) tool fabrication hours, (d) assembly hours, (e) wiring and plumbing hours, and (f) debugging.

machine. This problem can be illustrated by showing in Figure 10-2 typical percentage of completion curves for a series of different concurrent functions occurring in system development. These would include engineering, tooling, fabrication, assembly, wiring and plumbing, purchased goods deliveries and debugging. The different curves illustrate the inadequacy of bar charts, whether they are plotted out along labor hours or dollar expenditures.

Milestone achievement is perhaps the simplest method of determining and illustrating project status. Tables 10-1 and 10-2 are attempts to show a milestone sequence as it should occur in a normal machine development program. It is well to examine these in detail.

MONITORING THE PROJECT

Many assembly machine builders watch with mixed feelings the common letdown that occurs after receipt of a purchase order that involved months of effort on the part of the builder and the customer. It is as if the customer feels that they act as the father in the procreation of the machine, leaving the builder-mother to bear the months of labor in solitary if not splendid isolation.

Successful on-time project completion and, more importantly, a machine system with reliable sustained operational capability will be dependent on wise use of the time between order and machine assembly. *Most assembly systems are made or broken during the initial engineering phase of the project.* It is essential not only that the customer participate fully in the engineering phase, but also that the customer recognize that on-time delivery is dependent on the customer meeting milestones necessary for the builder to stay on schedule.

Tables 10-1 and 10-2 are designed to show the sequence and the interrelationship of the various events which builder and user must accomplish if the project is to remain on time.

Setting Project Milestones

Note again that Tables 10-1 and 10-2 are drawn in reverse order. Completion is on top, initial steps are on the bottom. This inversion is intended to show that assembly machine projects

TABLE 10-1. BUILDER MILESTONES

WEEKS BEFORE SHIPMENT	BUILDER				
	ENGINEERING SHIPMENT DATE	ELECTRICAL PNEUMATIC	FABRICATION & PURCHASE	ASSEMBLY	DEBUGGING
1	Pick up debugging changes	Complete temporary wiring		Paint, guard & crate	
2					Customer acceptance run
3					
4	Debug	Debug	Refabricate parts indicated in debugging		Preliminary visit debug system
5				Install standard guards & commercial units	
6	Pick up changes & errors in assembly	Wire and plumb	Complete fixtures		
7					
8	Complete assembly drawings & B/M	Wire panel		Install tooling stations & hoppers	Debug hoppers & individual stations
9			Make test fixtures		
10		Install panel box & ducts	Complete hoppers & fabricate tooling		
11				Erect changes	
12	Detail & release tooling				
13			Order weldmnets & castings		
14					
15					
16	Make formal layout drawings		Verify standard parts availability		
17					
18	Draw formal plan view-assign space to individual station				
19			Review vendor response		
20	Design fixture				
21					Develop & test in mock up state unique requirements of machine
22					
23					
24			Order feeders & commercial units as defined		
25					
26	Make concept layouts	Purchase motors & control equipment			Develop feeder concepts & fixture configuration
27					

TABLE 10-2. CUSTOMER MILESTONES

CUSTOMER			
CONTRACT OR PURCHASE ORDER	SAMPLE PARTS	SAFETY & I.E.	SHIPPING & RECEIVING
			Specify actual shipping date
Complete change notices			
			Notify riggers & truckers of probable & determine availability of rigging facilities
		Visit builder to review guards & operator positions	
	Ship final samples for acceptance run		
Review wiring, lubrication & hydraulic layouts		Verify actual service & utility requirements	
	Ship parts for debugging		
Consider visits to critical component vendors, feeders, leasers, etc			
Review layout drawings	Forward small quantities of production samples for comparison with older samples		
Assure that component drawings at builder are latest changes		Visit builder to review special guarding needs & review operator work stations & maintenance accessibility	
Assure builder has released long lead item purchase orders			
		Provide builder with latest info on physical area in which machine is to be located	
			Discuss any limitations concerning transporation & building access
		Review all pertinent I.E. & safety goals	
	Ship suffcient samples for hopper & work station development		
Review with builder actual status of purchase order			

must be built up from a basic foundation. The tables do not imply that any given delivery is six
months long. They do indicate that once the
builder's talent, time, and plant capacity are available (often many months after the order is
placed) machines of medium complexity can be
made ready for acceptance runs in approximately
six months of concentrated effort.

If the events listed do not occur in the sequence
and in the time spans indicated, the system will
probably be late. It is almost impossible to significantly reduce these lead times. The dates
shown are maximal times or in other words are
the last possible times for completion of any
phase and still maintain schedule. These tables
also assume that no major product changes occur in the last four months of project. In today's
world, that is a major assumption.

Communications and Changing Requirements

Among the various themes that seem to preoc-
cupy industrial management before disappearing
into an endless black void, somewhere between
the time when Zero Defects and later Value
Analysis promised to cure all of our problems,
manufacturing management was inundated with
the need to "Communicate." In assembly system
development, there is a tremendous need to communicate.

Far too often the communication between user
and builder tends to be horizontal, while at the
same time there is a lack of internal communication within large user organizations. It is important for the customer to recognize that since most
builder plants are proportionately quite small and
often have loose organizational structure, communications within the builder's plant occurs almost by osmosis. Almost every builder's employee
usually has a comprehensive if not detailed knowledge of what is going on at his or her own facility. *This ease of internal communication leads
builder personnel to assume internal communications go on with equal ease at the user facility.* A
tool design engineer discusses a point with a customer's manufacturing engineer and immediately
assumes that everyone in the user facility now
knows what is occurring and why. It is imperative that a customer technical and contractual

spokesman be delegated at time of order and that
some means of disseminating reports and questions be focused through these spokesmen inside
the user facility. It is particularly important where
manufacturing engineering and/or product designers are remote from the operating site or from
each other, or where capital equipment purchasing is done remotely from the operating site.

The Delivery Impact of Changes

A common source of customer misunderstanding is the fact that major changes in the
scope or requirements of an assembly machine
project almost invariably means that delivery is
affected to some degree. The builder too often assumes that the customer recognizes this point
and in turn neglects to document this impact, until a succession of changes entirely vitiates the
original delivery date projection. *Every purchase
change notice acknowledgment should include a
delivery date impact statement, so that delivery
promises continuously reflect actual status of the
order.* In some cases, the cost of delayed delivery
and the need for interim production facilities and
resultant higher unit costs may overrule the
smaller benefits of some proposed change.
Changes unfortunately do not seem to gain the
overall management attention devoted to the
original purchase. All members of the customer
team must be aware of the cumulative effect of
a series of small changes on machine deliveries.
Ford Motor Company has developed very functional systems to handle requested modifications
to a purchase order, in which proposed benefits
of the change must be compared to overall impact of delayed delivery before any final decision
for the proposed change is reached.

Using the Engineering Review Wisely

Chapter 11 outlines some of the formal reviews
that occur in the course of a machine development program. These formal reviews mark major steps in the development of the machine, but
in another sense engineering review is a continuous ongoing procedure wherein the fundamental decisions are made that ultimately determine
the net production capability, the ease of maintenance, and efficiency of ergonomic interface of
operator and machine.

The role of the customer must be that of the devil's advocate, during the engineering phase of the project. The customer must constantly question what provision has been made in the station designs for machine failure, what means have been made to facilitate the clearing of jams, what access has been provided for maintenance. *It is not sufficient that the design of machine and controls and tooling provide the means to make the machine work. It must also provide for all of the conceivable modes of failure by including the means for prompt restoration to service.*

The Human Factors. Machines are not run by engineers. They are run by operators who through economic necessity endure, or by personal preference enjoy, the repetitive nature of production work. The mental attitude of these workers is critical to obtaining high levels of net production. Assembly machines are usually not true automation. They can segregate defective assemblies, they can lock out subsequent operations after detecting a failure, but they usually cannot correct defects, or restore defective conditions to normal operating modes. Human intervention is required to assess and correct defective conditions. Easy access to the machine is essential if the operator is not to feel a conscious or subconscious frustration in trying to restore the machines to operation. Maintenance people will avoid or be slow to work on machines that cause skinned knuckles, ripped and stained clothes, and strained muscles.

Customers must realize that tool designers have a tendency to be problem solvers not problem seekers. They judge their efforts in the light of the functionality of their design to accomplish the specified task. There is something repellant to their creative (almost artistic) nature that wants to reject any possibility that their design might not function. The user must ensure in the design reviews that this tendency is overcome. *Machine design must provide for rapid restoration to operation in the event of machine or product assembly failure.*

Safety Considerations. In Chapter 9 we reviewed many of the governmental, legislative, and regulatory restrictions imposed on assembly machine users. From an operational standpoint, however, it is the design of physical and sound guarding imposed by the customer that will have significant impact on machine productivity levels.

As previously stated, the customer, or machine user, has the responsibility for playing devil's advocate when reviewing preliminary plans regarding the machine's capability for handling normal variations in piece parts. It is his or her equal duty to ask what design features have been provided to clear jams in an absolute minimum time.

An assembly machine operating at 40 assemblies per minute, which is assembling 10 parts, will process 24,000 discrete component parts per hour. The probability of finding one or more deformed or defective parts or foreign material among these 24,000 parts is very high. Not only must these 24,000 discrete parts be considered in themselves, but also other previously fed parts must be capable of receiving or mating with subsequently fed components. Add to this the variables that occur in joining stations and the thousands of cam rotations, solenoid operations, limit switch and relay actuations, and transfer device movements inherent in machine operations, and it is easy to see that, in the example cited above, there easily may be over 100,000 identifiable opportunities per hour for machine failure or stoppage.

With these factors in mind, it becomes obvious why it is so important in designing an assembly machine to devote as much attention to the ability of the assembly system operator to clear jams rapidly, as it is to handle good parts in normal operations. It would not be considered unusual for at least 10 of the 24,000 parts just mentioned to be so far out of tolerance as to cause some disruption. If the operator can clear each of these jams in an average of 10 seconds, less than 3% of the gross production capacity will be lost. If, however, it takes an average of one minute to clear the problem, 16% of production will be lost. For this reason every effort must be made in designing assembly machines to provide for the rapid removal of jammed parts or defective assemblies by machine attendants without the use of tools. Some of the design solutions to this problems are:

- Logic or memory-type controls that permit defective or incomplete assemblies to be bypassed. This approach often is deceptively attractive and is practical only if salvage or scrapping of incomplete assemblies is inexpensive, or the jam or hesitation which caused the failure to feed usually self-clears and the normal operation functions at the next cycle.

- Positive withdrawal, if at all possible, of all tooling prior to machine index. Rigidly cantilevered tooling can cause severe jams requiring disassembly of the machine or tooling. This, of course, is an intolerable situation.
- Reducing each operation on the machine to the simple functions. For instance, avoid feeding two parts into the same station or in screwdriving, separate screw feeding from screwdriving.
- Use simple mechanical actuations in the machine design wherever possible.
- Provide purge gates and quick-dump features on feeders and rails.
- Provide safe manual access for hand feeding parts into the machine in the event of feeder failure, or when individual stations are down for maintenance.

These basic design principles were receiving wide recognition by users and builders of automatic assembly machines before the advent of OSHA and the concurrent growth of product liability lawsuits. Now, however, guards for operator safety and hearing protection have become the order of the day, and the impact on machine productivity has been horrendous. The ability of the operator to restore an assembly machine to full production quickly is severely impeded when it is necessary to remove guards to obtain access to feeders and transfer units. Often such guarding is a last-minute addition initiated by safety engineers not responsible for production rates. Designing assembly machines to provide for operator safety as well as productivity must be considered in the very earliest stages of design.

Noise guarding and its related impact on productivity is a sore point with both assembly machine builders and users alike. Noise from assembly machines usually comes from three areas: fluid power components, part feeders, and bell-like component piece parts. The basic thrust to date in guarding against such noise has been containment through the use of physical barriers

(Figure 10-3). These barriers, however, restrict visual and manual access.

Project engineers and machine designers should try to eliminate noise in every way possible during the design phase of assembly machines. Cam actuation can cut both noise levels and energy costs significantly. The rapid release of compressed air from pneumatic components, on the other hand, is a basic source of industrial noise in many factories. It should be avoided wherever possible. Either avoid the use of compressed air or provide suitable mufflers.

Large hoppers for part feeding can through their very operation cause significant noise. Many unnecessarily large hoppers are used, not because of any functional requirement but because of inflexible procurement practices regarding piece part storage capacity. Use of minimum feeder bowl diameters can limit the noise problem and may eliminate the necessity for noise enclosures. Tremendous advances in part feeding technology in recent years have produced a new family of centrifugal rotary parts feeders and linear belt feeders (Figure 10-4) specifically designed to eliminate or reduce the necessity for sound enclosures. It probably is too early to tell whether or not these feeders will have the durability or range of applications of earlier vibratory feeders. Regarding linear feeders, it is particularly disturbing to note the tendency of builders of such

Fig. 10-3. Sound enclosures on a large vibratory feeder defy easy manual and visual access and reduce productivity.

equipment to increase feeder sizes. This inhibits easy access to the assembly machine.

The desire to place physical barriers in front of moving parts of assembly machines has brought additional woes to engineers and managers who want to improve productivity. To date much of the emphasis has been on fixed physical restrictions to fixture and station access. Such barriers are burdensome to operators and often are removed or inhibited by machine attendants in the field, thereby leaving users open to OSHA penalties and compensation claims, and machine builders vulnerable to liability suits.

Two approaches to more realistic safety systems are often ignored in initial design activities. Various forms of radiation guarding, which use the Doppler effect or changes in capacitance or induction, have been adapted to various assembly machines with differing degrees of success (Figure 10-5). One problem with these systems, however, is the necessity to detune the system so that normal machine operations do not trigger the safety devices. It is important to determine if there are any dead areas by the use of field-strength meters. For the most part, such systems are rarely used in new machinery.

Light curtains (Figure 10-6), which originated in Europe, have been widely accepted in the United States, and offer excellent solutions to safety problems. They have minimal impact on

Fig. 10-5. Round tubes are antennas causing a radiated field capable of sensing human presence. Note detuning plates or shields near operator positions.

machine productivity and offer uniform sensitivity throughout the area of coverage and offer excellent access to machine tooling. Increasingly, however, barrier interlocked guards are preferred by international organizations (Figure 10-7).

Maintenance Considerations. At the cost of boredom, it is necessary to emphasize again the importance in the design review of critiquing proposed system and station designs in the light of possible machine maintenance problems. Access to index units, lubrication systems, and wiring terminal strips is critical. *A theoretical goal during design should be to provide maintenance access to the basic machine and control systems without removing or disturbing tooling stations.*

Consideration of Probable Downtime and Restoration to Operation. Maintenance failures in a well-designed assembly machine should not be commonplace, and a well-designed preventive

Fig. 10-4. Linear feeders use belt chain or straight line vibratory tracks to move component parts through selection and orienting devices.

Fig. 10-6. A light curtain is an encoded light beam transmitted, reflected, and received by a sensing device capable of detecting minute penetration of the screen.

maintenance program should prevent minor problems from becoming major breakdowns.

Machine stoppage, however, due to jams caused by foreign material or defective components is a constant problem. There are practical limits to the qualification of incoming components. Sorting feeders can detect many of the problems on external surfaces but are usually ineffective in determining part cavity quality levels. A pin might not go in the hole because the pin is bad, but the chances are equally good that the hole might be defective or improperly located. In Chapter 4 we examined the impact of downtime on machine productivity. Once the machine system is sound, the part quality and consistency is as high as practical, and the machine properly installed, there remains only one way to increase productivity and that is to ensure that jams and stoppages can be cleared easily and readily so that downtime is minimal.

If the problem can be operator cured, rather than calling on maintenance, downtime will be significantly reduced. If the problem can be readily identified through the use of visual display panels, we have aided productivity. If problems can be cleared by hand rather the use of hand tools, we have eliminated not only extended downtime, but possible damage or misalignment of the machine or tooling stations. If creative guarding, such as light curtains, permits safe, easy access, we will greatly increase production. All of these elements should be considered in the concept and layout reviews.

Preparing the Operators

"Operators" is here intended to be a broad term including all of the people who directly or indirectly will determine ultimate machine productivity.

An assembly machine, no matter how large, sophisticated, or costly, is nothing but a manufacturing tool. It does not run itself. It must be run

Fig. 10-7. European barrier guarding is becoming widely accepted in the United States and Asia.

and run aggressively and wisely. It has no motivation and little, if any, corrective abilities.

Its introduction will often bring unease or fear to the plant workers as they find their jobs threatened or eliminated by "automation." Failure to realize the implications of the above will delay machine startup and may permanently cripple the machine is productive potential.

Site Preparation. In the last quarter of a century, it has been the author's experience that 20–30% of all assembly machines shipped arrive at plants unprepared to receive them. It is a major failure not to involve industrial engineers and plant engineers in the logistical problems of machine installation as soon as the layout drawings are complete.

Among the more common installation problems are:

- Failure to check machine size against entry door size, elevator capacity, and internal aisles and doors.
- Ensuring that sufficient compressed air of clean dry condition is available.
- Ensuring that proper staging areas are available for the incoming components and completed assemblies.

If these seem trivial considerations, ones that most industrial engineers consider routine, remember that many assembly machines are placed in plants or plant areas where no significant capital equipment exists, and often where plant engineering skills might be oriented to simple maintenance, heat, light and power, and security operations. The assembly machine under consideration may be the first significant piece of machinery introduced since the original plant construction.

People Preparation. It is very common to have specification procurement and design review of assembly machines done by staff rather than line personnel. This may be done at a site far remote from the actual production plant. It is a mistake, particularly when working on an existing product, not to have line operators participate during all of the phases of machine development. This participation will be discussed in the next chapter.

SUMMARY

The quotation, selection, and procurement practices used in obtaining an assembly system will determine probable production levels.

The design review is the critical stage, however, when errors or omission at time of procurement may be rectified. Additionally, design review will determine whether the machine is to be a theoretical or practical solution to assembly and quality problems.

Chapter 11
Machine Acceptance and Installation

INTRODUCTION

The acceptance procedure in approving automatic assembly equipment for shipment, installation and start-up is composed of a series of small vital steps beginning with vendor selection and ending only on turning over the equipment for ramp up to full production operation. Those having the responsibility for making acceptance decisions should understand their obligations, their responsibilities. and their prerogatives.

Equipment designers are technically oriented; they are solution seekers. The mentality that makes them adept at their creative job may mean they are poor in understanding the operational, maintenance, and even political environment into which any new assembly system is to be placed. The buyer group assigned to the purchase and acceptance task must review proposed technical solutions in the light of their probable operational environment. *The role of the approver is not to design the equipment but to determine the practical realistic nature of the offered solution.*

The prerogatives as acceptor must be exercised in a morally sound, equitable manner. Development time for an assembly system is costly. If it goes beyond the originally projected time, costs multiply for the builder, and profit is lost forever to the user. Development will proceed most smoothly if possible problem areas can be identified, isolated, and corrected at the appropriate point in time. The most critical of these areas are explored in this chapter. Simultaneous engineering adds another dimension of difficulty, since downstream product changes may invalidate prior tooling process and equipment decisions.

Contractual payment terms, discussed in Chapter 10, may weaken or strengthen the hand of the approver. The one holding the money is in a stronger position.

The comments in this chapter are based in part on the fact that *most buyers are rarely in a position at time of machine buy off to live up to their own formal acceptance criteria developed in contract negotiations.* Few machines ever meet the legalistically worded acceptance criteria included in many formal purchase orders. This topic is explored in detail in this chapter.

Time, production schedules, and sample costs will be basic factors in the ultimate decision to accept the machine from the builder and turn the machine over to production. This sometimes traumatic decision can often be simplified and shortened by the following suggested steps.

ENGINEERING APPROVAL

Assembly machine acceptance is too often visualized as a single momentous decision at one point in time, when in fact it is a series of small decisions, each one of which increasingly limits the flexibility of the approver in making subsequent decisions.

In fact, by selecting a specific system vendor's proposal for purchase, one has already substantially narrowed available options. Too often buyers, at this point, may feel they have made their choice and placed themselves in the hands of the builder and thus may assume a relatively passive role. Such a passive attitude will generally lead to late deliveries, vague acceptances, and poor performance throughout the life of the equipment.

Before looking at the user's role in the development of the assembly system, it is well to look at what happens at the system builder's facility once a purchase order is received. There is a switch from the sales function to that of design and build. The formal proposal and its matching purchase order, no matter how lengthy, are but brief statements of months of discussions, conferences, and informal communications. Salespeople, to be successful, must be sensitive to the aspirations, motivations, and operating environment of the potential user. An assembly system salesman is selling a problem solver, a solution to a complex multi-faceted manufacturing problem. Once an order is received, however, the system builder's role becomes that of a system developer, and different skills and different people are now applied to the job. The basic builder's thrust is now toward a technical solution of the problem as a machine designer sees it through the words of the proposal and purchase order. It is necessary that the user be certain that the problem the machine designer or system integrator sees is one of actual current and future production and operating environment requirements, not just the words of the proposal and order.

Customers who have the responsibility for engineering approval of any proposed assembly system must not be content when a specific technical concept is established to transfer or join or inspect the product as it is assembled. The phrase "excellent in concept, poor in execution" reflects a chronic problem in assembly machine system development. An excellent concept must be executed in a way and with materials compatible with high cyclic operational rates, remote plant locations, and limited operating skills. It often happens that theoretically excellent concepts cannot be well executed and may have to be scrapped in favor of less exotic but more durable solutions.

It is the purchaser's duty to determine that technical concepts for each functional station are well executed before releasing engineering for fabrication. *It is the customer's right to insist that the engineering solution be operationally viable.*

The Model Making Approach

Many of the normal assembly requirements in part transfer, joining, and gauging can be handled with straightforward, commercially available and proven devices. Each assembly project, however, generally contains one or more unique problems, problems specific to that assembly. Far too often it seems that one attractive way to tackle these problems is by a model making, empirical approach to problems of fixturing, part selection, transfer, or joining. Simple or complex mock-ups are used to determine feasibility of suggested approaches. The feeling, often, is that there is nothing better to determine the practicality of proposed part handling than a hands-on solution.

There are, however, several pitfalls to the model making approach to system design that the machine approver must be aware of:

- Model making may be a lazy way to solve a problem fully capable of a pure engineering solution without recourse to model making. Developments in analytical and simulation software make this statement even more true.
- It is extremely difficult to copy dimensions on models accurately. These dimensions often are empirically devised and yet functionally critical.
- Copying of dimensions on accepted prototypes or models for formal engineering documentation is often left to the lowest levels in the drafting room and often to beginners who do not fully understand the functional requirements incorporated in the model.

- If the mock-up works well, there is a strong temptation to incorporate the actual mock-up in the machine instead of duplicating it from engineering drawings to known dimensions and tolerances with properly engineered materials.
- If stations built on a model making basis are incorporated into a production machine, one can almost be assured of difficulty in replacing worn parts. This problem is compounded if a series of duplicate machines are required for increasing production volumes. It becomes virtually impossible to maintain a stock of replacement parts which are fully interchangeable.

The *true role of model making should be to prove the feasibility of proposed concepts*. The degree of sophistication of each conceptual mock-up will vary, but it should incorporate actuation motions and cyclic rates identical to the proposed production cyclic rate. Model making is an excellent development tool, but should not substitute for complete engineering documentation. Documentation aids in future maintenance requirements. Documentation is essential to any future rapid changeover requirements for running tooling changes particularly when tooling modification must be done at a site different from the operating location. Running modifications on operating machines are increasingly the norm.

In viewing a model or prototype development for further engineering, the acceptance team should be forthright about their feelings concerning the suitability or lack of suitability of the proposed concept. Modification is still easy and inexpensive at this time. (Figure 11-1.)

Formal Tool Design

The formal tool design effort should begin with certain essential elements. These must include at the very onset a reasonably accurate plan view of the machine and a detailed proposed fixture design.

The plan view of the proposed layout will serve several important functions. It is a space allocator defining the room available for peripheral equipment and the space available for operators to load parts and clear jams. It permits the as-

Fig. 11-1. In reviewing proposed station concepts, mock-ups are extremely valuable, but should not be considered a substitute for complete engineering drawings.

sembly system design group to discuss preliminary orders for purchased equipment, particularly vibratory feeders. Quite often, for example, when feeding complex parts, it may prove necessary in feeder development to change from clockwise to counterclockwise rotation. This may require a totally new overall system layout. It seems that with the advent of outboard tooling vibratory feeders tend to grow in size, a problem compounded by the addition of sound guarding. The trend to work cell configuration with buffer storage between machines is another example where original layouts or space allocation may prove invalid.

The plan view will remain in a constant state of flux as vendors of feeders, joining equipment, and inspection stations begin to respond to the system integrator with their initial layout drawings. At this point it is essential that peripheral tooling does not inhibit operator and maintenance access to the machine. Maintaining the accuracy of the overall layout is greatly simplified by CAD, but the discipline of maintaining its accuracy is vital.

The initial fixture concept may also require continuous modification as each station layout is developed in accurate, scaled drawings. Access for transfer fingers, additional locating units, strippers, and hold downs may force continuing revision of the fixture and even the proposed sequence of operations.

At this early design stage, the *customer team must constantly guard access room* for the clearing of jams, minor adjustments, and actual loading of parts to the work-holding fixtures or to operator-assist units. It is particularly important that the purchasing group visualize the prestaging of component parts for the various part feeders and, in particular, ensure sufficient room for the ejection of completed components. The ejection process on most machines is such a minor technical matter, compared to the part feeding and joining operations, that it often becomes a neglected item in machine evolution. Part feeders and joining stations, prior and subsequent to ejection of the completed assembly, tend to encroach on the ejection station. The builder often initially plans to eject completed assemblies to a simple chute without any real consideration of what is to subsequently happen to the completed assemblies. These potential problems give rise to

two other vital uses of the plan view as soon as it is available. Customer personnel with responsibility for industrial engineering and plant safety should receive copies immediately of each modification so that installation problems and material flow and guarding requirements be faced up to early in the layout stages. The plant location of existing columns, services, aisles, and ceiling heights may be critically important in the machine layout, particularly so in integrated plants. (Figure 11-2.)

The natural tendency of stations to grow in size and number during engineering development is particularly important if a dial type chassis has been selected for the system. Until we convert to some new type of measurement, circles will consist of 360°. Indexers in most types of rotary machines are notoriously inflexible as to the number of available stops or index stations. Linear machines offer a great deal more flexibility in overall system layout if the machine becomes crowded as layout work progresses.

Ongoing station development and implementation problems in such overall system development may indicate a real need to modify the proposal sequence of operations. Revised operational sequences may force fixture design modi-

Fig. 11-2. Industrial engineers should review proposed layouts as early as possible. Access for maintenance and material flow is critical to successful machine operation.

fications. The plan view and fixture design must remain fluid until all station design is fairly well established.

In this evolutionary conceptual design period the review team should also watch fixture development carefully. Ideally, the fixture should be as simple as possible. Fixture costs on synchronized machines may be as high as 15–20% of overall machine costs, and 10–15% is a common range. In nonsynchronous machines, fixture and pallet costs are very major considerations.

There are several pitfalls that should be avoided in fixture design. In assembly machines, the requirements are somewhat different from metal cutting machines. A metal cutting machine usually clamps the work piece securely. If for some reason a fixture is empty, the cutting tools generally are cutting air and the chances of damage to tooling are minimal. In assembly machinery, to the degree that the design will allow, the fixtures should be simple nests, capable of transporting the product from operation to operation and able to withstand joining pressures and torques. (Figure 11-3.) They ideally should not

clamp the work pieces, since in many cases there may be valid reasons to remove a partially completed assembly from a fixture and/or replace it with another subassembly anywhere in the machine sequence. The necessity for this capability is discussed elsewhere in this chapter.

It is also good design to avoid the incorporation of specialized locating features in the fixture, wherever practical, and instead attempt to provide critical locational devices only in those stations where accurate location is important. It is far easier to maintain the accuracy of one locating device required for some insertion, gauging, or joining operation as an integral part of that station, than to maintain that capability in 40, 50, 60, or more fixture nests. (Figure 11-4.)

In addition to locators, it is also most important to avoid specialized joining devices in each individual fixture. Riveting mandrels and lower welding electrodes and similar devices, if at all practical, should not be built into fixtures. The use of riveting mandrels and lower welding electrodes as part of the fixture design will usually simplify the builder's design task since they may

Fig. 11-3. The fixture nest shown on the left is extremely simple and will require little maintenance. The fixture on the right will require more maintenance. The complexity and weight of the fixture on the right, necessary because of product design, adds to inertial loads on index and will require greater maintenance.

Fig. 11-4. A critical location is established in the work station, greatly simplifying nest design.

even serve as a partial or complete fixture nest. From an operational standpoint, however, they become maintenance nightmares requiring exact dimensional conformity and as they become damaged or worn, are difficult to identify and isolate. Depending on the control technology used, failure or wear in one or two such fixtures may bring a complete machine system to a halt. If these perishable joining tools are incorporated into the station design, they must be easily adjusted for height or location. Failure or wear must be easy to spot, isolate, and repair.

From a builder's standpoint, it is probably very inexpensive to incorporate these components in the fixture. Wear will not be notable or significant during the machine tryout period. There is often an attempt to rationalize such design practices in terms of spreading wear over a large number of stations. The logic is fallacious. If a welding electrode can take 30,000 welds before arcing and weld pressure take their toll, there is no difference in replacing one lower electrode 50 times after each 30,000 cycles or replacing 50 lower electrodes after 1,500,000 cycles. If these mandrels or electrodes or precise locators in the fixtures are not adjustable, they may prove to be more costly than those designed for specific joining stations which are often adjustable, and/or reusable.

The design review and preliminary acceptance of station concepts will be one that most customers

find difficult. Unless they have had a broad background in mechanized assembly, they will usually defer to the machine designer. Such deference to greater experience is not without cause, particularly when the acceptance team is composed of electrical or computer science graduates with modest backgrounds in machine design.

In previous chapters discussing part feeding, transfer, joining, and inspection, we have discussed many of the elements that go into good functional design. At this point in the design review, the customer must insist on seeing design provision to clear insertion failures or defective assemblies, hopefully, by operators without the use of hand tools and not, of necessity, by maintenance personnel.

In this review of formal layout drawings, provision for the avoidance of potential dead jams is important to prevent machine or tooling damage. Compensation features are critical in station design. Ideally, all parts insertion should be done with spring loading or light air pressure so that failure to properly insert a component will be compensated for by spring or air compression. Withdrawal of tooling prior to index, however, should be positive. Wherever possible, tooling should be with drawn at each cycle of the machine. The use of rigidly cantilevered tooling, locators, and strippers should be avoided over or under indexing fixtures.

Some years ago engineering documentation for special machinery was sketchy at best and many customers neither requested nor expected tooling detail part drawings. It was common practice to send out only a simple set of layout prints at initial installation. Whatever existed in the way of formal tooling drawings usually remained at the builder's plant. Ford Motor Company was a significant major exception with their early insistence on the use of Ford drawing standards and the decimalized inch. By the 1980's, full documentation of tooling was usually mandatory by most customers and user drafting standards were commonplace. In today's environment, broadly accepted CAD software, such as Autocad, has replaced corporate standards. Electronic transfer of engineering from builder to customer is the norm.

Conformity to customer's engineering standards was and remains a very significant cost element in most assembly machine projects. These costs include excessive documentation beyond that necessary to build and maintain the machine. They may force deviation from the builder's normal practices with resulting losses in efficiency and delays in obtaining purchased components other than those normally used by the builder. These extra costs and delivery delays are worth discussion and evaluation before final discussions are made.

Drafting Standards. Most users, and particularly those with large inventories of specialized machine tools, had insisted on drawing to their standards, preferably on their paper and particularly with their drawing identification numbers, their detail numbering formats, and their bill of material format. Such a request seems eminently reasonable unless one takes a second look at the problems of conforming to such requests.

The increasing use of computer aided drafting and design and the dominance of two or three major software systems is providing significant relief to the problems described in the above paragraphs.

A large assembly machine builder may have as many as 30–50 assembly machine projects under way concurrently. These may represent 20–35 different customers, and each machine may have (excluding screws, nuts, etc.) several thousand different component parts. Depending on the degree of standardization employed by the builder, the machine may consist of 1000 or more standard commercial components and several hundred unique tooling details and purchased parts. Tracking of these parts is best done by computer, but the software on most computers demands standardization in part description. The builder's normal bill of material procedure should be compatible with his or her computerized production control system, but the whole system is severely crippled when customers insist on their bill of material format.

Many bill of material formats specified by customers evolved years ago to satisfy the relatively simple requirements of in-house engineering of dies, molds, and fixtures. These bills of material are often inadequate for the sophistication required to monitor and manage a large assembly machine program. There should be a recognition that the builders are the ones who build the machine, and insistence that they deviate from their normal production control practice can only adversely affect their operation.

Too often the total engineering review activity has little to do with functional or operational characteristics of the proposed machine, but an inordinate amount of time on such review is devoted to drafting standards and bills of material numbering systems.

Interchangeability. In any complex assembly machine, there will be standard builder components, purchased parts, and tooling details which because of their application will have limited life. Such parts may be of necessity fragile, or because of the high cyclic rates and inertial loads common to intermittent motion machines, may be considered perishable or semi-perishable. Later on in this chapter we will discuss spare part kits as part of the procurement effort. In the engineering review, the user should look for several tendencies that can lead to lack of interchangeability and resulting significant maintenance downtime.

The first tendency to avoid is design which makes reference to "fit on assembly." There will be some parts of most assembly machines which can be justifiably so dimensioned. Vibratory parts feeders and tracks are usually as much art as science and hence, dimensions on such feeders are at best broad references. Major mounting brackets, weldments, or castings used to mount tooling stations may have little functional need for close dimensional tolerances. As long as they are stable and hold the tooling firmly and squarely they are adequate. Wiring harnesses and cable and piping runs usually need not be accurately dimensioned. These are usually installed after mechanical assembly on an ad hoc basis. With these exceptions, notes or drawings which state "fit on assembly" usually indicate a lazy attitude toward engineering. This note in turn will lead to undue maintenance delays if used on tooling subject to accidental damage or rapid wear.

A second check in the design review is to ensure that commercially purchased components are identified as to the original manufacturer and, where applicable, to the OEM's standard catalog identification numbers. They should not be identified by builder's or customer's numbers (unless it is a user's commodity identification number), since it inevitably means unnecessary ordering problems, price mark-ups and delivery delays.

Many users have established standard lists of preferred or mandated commercial components that must be incorporated into machine design. The rationale given for this is highly logical: a reduction in the number of stocked maintenance parts and a reduction in downtime because of increased maintenance part availability.

In practice, however, the supposed advantages are not always available. Recommended commercial components may not be adaptable to the builder's machine design. Surprisingly, very often such recommended lists include obsolete or obsolescent parts. They may prohibit the builder from utilizing the latest component designs. Too often, unfortunately, they reflect personal relationships at corporate purchasing or engineering locations, without full regard to availability, cost, and service on such components at the builder site or in the region where the machine is to be located.

Computer-Aided Drafting. The rapid advent of computer-aided drafting (CAD) will probably mean significant adjustments in attitudes toward customer drafting standards. (Figure 11-5.)

Much of the drafting effort on assembly machines tends to be repetitive, particularly where modular machine systems are utilized. Many stations combine a few tooling components, such as pick-up fingers with standard transfer units. These standard sub-assemblies and standard component parts can be stored in the software packages associated with the particular digitizer or computer-controlled plotter used by the machine builder.

In order to take advantage of the reduced drawing times such systems offer together with reduced costs and lead times, it may be necessary to waive some or all of the user drafting standards in favor of standardized drawings produced by the builder's CAD system.

It should not be forgotten that the primary user for all tooling drawings will be the machine builder's fabricating and assembly departments. Deviation from the builder's own operating procedures and standards can lead to costly delays in the project.

Analytical Design. As this book is written, it has become increasingly clear that the advent of an-

Fig. 11-5. Computer-aided drafting can dramatically reduce engineering times on assembly machinery. This potential savings in time and money may be lost by undue customer restrictions imposed by their corporate drafting standards.

alytical software designed for PC's and work stations is causing a revolution in design practices. The historical sequence of concept design, formal layouts, detail drawings and bills of material followed by analysis is rapidly being replaced with appropriate CAD software which permits design to be done in an interactive concept development/analysis mode with formal design and bill of material preparation becoming byproducts of the concept development.

Implications of Customer Approval

What is the true role of customer approval? It is a temptation to slip into the legal implications of a step-by-step, sometimes grudging, concurrence with the assembly machine design, fabrication, assembly and debugging efforts. It is easy to consider the various approval steps as a gradual assumption by the customer of responsibility for the machine's ultimate production performance.

In one sense the customer approval (or lack of disapproval) of overall system layouts and station concepts does indicate concurrence with the design effort, but it should never be viewed by either party as acceptance of responsibility for performance of the system or any given station. *System design responsibility must remain with the builder until final approval of the machine.* It is important that progress payment terms do not, by their wording, imply acceptance, but instead acknowledgment of milestone achievement.

If this seems to be a game in semantics, it should not be. Twenty five percent (of the machine cost) due and payable upon completion and release of engineering is totally and radically different than 25% due and payable on customer acceptance of engineering.

In brief, customer "approval" of design engineering or integration software should reflect the judgment of the customer that the proposed design appears feasible, seems capable of high cyclic activity, and is maintainable in the operating environment in which the machine will ultimately be placed. Customers would be well advised to use the term "engineering review" rather than engineering approval. Their concurrence should not convey acceptance of responsibility for performance until such performance is a fact on the factory floor.

This calls for mature objectivity on the part of the review team. The review team must avoid the situation where industrial engineering, safety engineering, production managers, and others who may have been only incidentally involved in procurement of the machine now attempt to change the scope and intent of the project. *Many a builder has found that a customer engineering review visit often turns into a customer staff meeting during which the customer's lack of internal communications becomes embarrassingly apparent.*

In one sense the initial review of a mechanized assembly system can expose many previously overlooked managerial problems by the sharp focus of any engineering review.

DEBUGGING

It is overly simplistic to consider debugging as a single phase of machine development a few short days between the completion of machine assembly and shipment to the production plant. Would that this were so. In fact, debugging must start with the first day of tooling mock-ups and ideally continue throughout the life of a machine. If the Japanese excel anywhere it is in the continuous improvement of existing equipment. If debugging is conceived as proving the machine capability, it will only prove partially useful. Knowing the complete role and the true nature of debugging is critical to the second stage of machine acceptance. It is the complex art of turning the theoretically feasible into the realm of the functionally practical.

Debugging of an assembly machine is always directly related to sample product component quality and availability. Debugging is not reengineering. It is the process of adapting a sound machine and control design to accept the widest possible range of component parts variations.

Sample Availability

In Tables 10-1 and 10-2 guidelines of critical dates are given to ensure a project remains on schedule. Many of these dates are concerned with the availability of product samples. It has been mentioned earlier that there are many more assembly machines manufactured for new products than for cost reduction programs on existing products. Additionally, the trend toward simultaneous engineering makes sample availability far more difficult. Even when a machine is purchased to reduce costs on existing products, there may have to be major design or dimensional modifications made to component parts to facilitate assembly reliability, and there may be difficulty in obtaining samples of these modified parts.

It is extremely difficult (and implies a great deal of experience) for any builder to firmly quote a complex assembly job on the basis of part prints alone. Many simple jobs lend themselves to mechanized assembly. It is easy to visualize the behavior of parts such as washers, screws, and pins in part feeders; it is much more difficult to do this where component parts have complex or involved shapes. For this reason prototype sample availability is helpful at the time of proposal. It is not that the builder cannot visualize the appearance of the parts and the assembly itself, but rather that the prospective builder would like to observe the behavior of the components under the forces of gravity, inertia, and momentum. The location of proposed parting lines, sprues, gates, and mold numbers may also have direct impact on the proposed assembly sequence and tooling. (Figure 11-6.)

Once an order is placed, there is nothing more vital that a customer can do to help an assembly machine program run smoothly than be certain that samples are available in the quantity and conditions required for each phase of the project. A few prototypes may suffice for significant fixture, hopper, and escapement development; single-cavity molds and dies may be practical for initial hopper development and early machine construction and even early debugging. In the long run, however, significant quantities of actual production samples are required to complete sufficient debugging to permit shipment of the assembly system. Failure to do so will result in extended delays in reaching full production.

If lack of samples is a problem, early delivery of samples for trial runs before the builder's requested dates may prove equally troublesome. A large number of parts are required for hopper development. It is to be expected that many of these will be lost in the development of the parts feeders. Those samples returned to the builder by the feeder manufacturer with the feeder may have been fed and refed countless times in perfecting the feeder. These samples may be useful for preliminary station debugging and timing, but they are generally unsuitable for fully automatic trial runs and debugging. The parts used for hopper development often are tryout parts from new dies, molds, and tooling. They are usually ade-

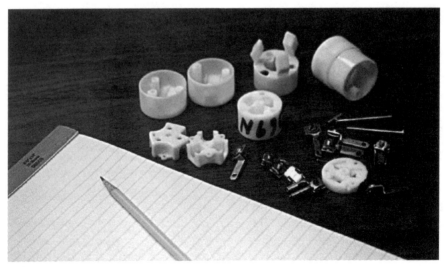

Fig. 11-6. Samples, even prototypes, are most helpful in preparing assembly machine proposals.

quate for substantial hopper development. If large numbers of such parts are available at the time of hopper development, there is a strong temptation to retain a portion of these parts for final debugging. Five or six months may elapse from the time the early samples are received by the builder and the time when debugging begins on a system in an automatic mode.

The final debugging should always be done with the newest production samples. These will reflect not only die, mold, and tooling modification on parts made within the builder's plant, but will also give a clear picture of the quality and consistency of vendor supplied components.

It is essential at this point that whenever multiple vendor sources for any component are utilized, samples from each vendor are included in the samples supplied for final debugging and acceptance runs. Additionally, no matter what the source, parts from each cavity or die should be supplied (and identified as to the specific tooling) whenever multiple fabrication tooling is utilized.

Purchased parts are most troublesome here, not because vendor quality is necessarily worse than that of components produced by the customer, but because deviation requests from prints or specifications are often funneled through purchasing, quality control and product design sections of the customer organization without any notification to the manufacturing engineering group responsible for the assembly machine.

Sample availability then is important in five areas: proposal preparation, initial concept and fixture design, feeder development, debugging, and, finally, acceptance runs at the builder's facility and on the production floor. The number of parts available and their resemblance to the final production parts become increasingly important at each successive step of the machine evolution. Delays in sample availability do not mean simple equivalent day-by-day delays in machine delivery, since machine, control, and feeder vendors must ship on a regular basis to survive economically. If parts for one project are not available, the builder of assembly systems of economic necessity must turn to projects where parts are available and complete these for shipment before returning to the original program.

The Nature of the Debugging Process

If the engineering on an assembly system is totally correct, and the component parts to be assembled are to specification and identical, there would be a modest need for debugging of automatic assembly machinery. This does happen, but is so rare that one should not expect it in their project.

Debugging is a normal component of machine development which should be a continuous ongoing activity throughout the product life cycle. A full understanding of its role and its nature and how it is best done will reduce the lead time to bring an assembly machine up to a profitable rate of productivity, and through a continuation of the debugging process, continue to improve the productivity of the machine until a point is reached where further debugging expenses are not matched by identifiable increases in productivity.

It is unfortunate that "debugging," a negative word which implies a short term activity, is so commonly used. "Continuous improvement activities" is not a euphemism. It is a statement of necessary action.

In several previous chapters we have discussed the real meaning of efficiency, net production, and gross production. By now the reader should readily see net production as that percentage of gross production which remains after deducting the product of resultant inefficient machine cycles and the average downtime. It is clear that reducing either the number of inefficient cycles of the machine or the length of the average downtime will result in greater net production. Sound engineering of the total system—chassis, fixtures, controls, and operating stations—will contribute to reducing the average downtime. Creative approaches to guarding will also be most beneficial.

Reduction in average downtime essentially will come in the engineering stages of the program and by proper operator education discussed later in this chapter. Debugging is intended to improve the other factors controlling net production—system and station efficiencies.

In previous chapters it has been pointed out that failure to assemble or join a part can result from failure of the system to coordinate activities; failure of the fixture to stage the parts properly; failure of the individual stations to perform their design function; inability of the particular component being assembled to be fed, oriented, transferred, or inserted because of its deficiencies; and, lastly, the inability of previously fed parts to receive subsequent components. The opportunities for any type of failure to occur when one considers the number of parts being fed per hour are indeed enormous. In engineering the system and its individual stations, the system designer tries to look at the parts being assembled and imagine the conceivable variations that the machine might be expected to handle. The designer will try to design into the machine the ability to accept parts with dimensional or material differences, and, if possible, incorporate features to segregate and eject defective components automatically. *The degree and extent of debugging necessary will be determined by the success the tool designer has in providing for acceptable product variation at the design stage.*

Debugging essentially consists of four steps:

- Observing the behavior of fixtures, feeders, transfer devices, and joining stations when exposed to a broad spectrum of component parts.
- Analyzing the types of failures that occur, and recording the frequency of these failures.
- Making a decision as to the nature of the required correction, and where a variety of options are suggested, choosing the correction that is most appropriate.
- Implementing the changes in the most expeditious way.

These deceptively simple steps become quite complex when the conditions in which debugging occur are considered. Debugging is a never ending process. It is not something that is an isolated phase of machine development. Debugging begins when fixture concepts are established. Additional debugging is done during mock-ups of unique stations on the machine. Extensive debugging is done during the development of feeders and discharge tracks, because, in manufacturing feeders, the processes of debugging are the main tools of development. There is little real difference between the model making approach to machine development and the debugging process.

For most people, however, debugging means the steps that are taken after a machine is assembled to bring its net production to acceptable levels. The initial trial runs of most assembly machines reveal little of its final potential. Initial timing and sequencing often must be constantly adjusted. The first time the fixtures are loaded with parts may reveal embarrassing clearance problems. As peripheral equipment is mated to the main machine, particularly those supplied by outside vendors, additional problems of interference are discovered. Correction of these problems hopefully are minor annoyances and are not true debugging. The actual debugging process consists of improving inefficient operation at a basically sound machine to acceptable operational levels. Debugging will be involved with the machine's capability to efficiently handle functionally acceptable variations in component piece parts. It should also include the ability of a machine to function as a machine under continuous production levels in the environment in which it will be operated. (Figure 11-7.)

In the first few hours of machine cycling, improvement in productivity is often dramatic. The isolation and correction of one problem, however, often exposes another. The first few hours are generally concerned with timing and mechanical adjustments. The removal of sharp edges in tracks, fingers and fixtures and jam points in feeders as fresh new production samples are fed into the machine will prove necessary. Production goes up initially in significant steps, but each new plateau of production efficiency becomes more difficult to achieve as the original obvious problems are isolated and corrected.

Some experienced builders take a very systematic approach to debugging, running each station on the machine in isolation, starting from the first station on the machine. This methodical approach has much to recommend it. There is also much to be said for an initial rapid, preliminary examination of each station individually and the entire machine system under power before going back to the systematic examination of each individual station. A quick preliminary overview of the entire machine will give a good idea of the extent, the nature, and priorities of problems to be corrected in the debugging phase.

If there is not an early attempt to operate all

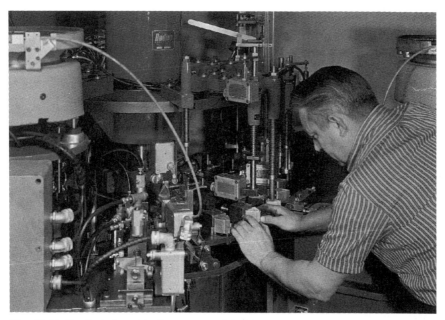

Fig. 11-7. Analysis is critical to successful debugging. In the final stages, it may require days to discover and correct a simple problem.

stations both individually and in coordination with other stations, one will go systematically to one station at a time and solve every problem consecutively. If there is a rapid preliminary trial of all stations, debugging efforts can often go on at several stations concurrently. Once the major problem areas are identified, a plan of action can be made that will complete debugging in the shortest possible time. It is most frustrating to go through 10, 15, or 20 stations only to find that the next station has problems that can be cured only through the purchase of some new components with extensive lead times. A quick preliminary run through will identify these stations, leaving the debugging of stations not dependent on outside resources to later in the project.

In Tables 10-1 and 10-2, the last four weeks prior to shipment are reserved for debugging. This period is a reasonable period, on average, for medium sized assembly systems. This four-week scheduled period may begin to shrink as unforeseen problems occur in obtaining try-out samples, product modifications are incorporated into the machine design, and deliveries of major commercial components for the machine fall behind schedule. Pressure begins to mount, both on

the builder and the customer's purchase team, as the promised delivery date draws near and the machine is not yet ready for shipment. The time required for analytical, objective study during debugging may seem to be one of inactivity on the part of the builder. Political pressures inside the customer's own organization can lead to a demand (no matter how politely worded) to "do something." Doing something physical rather than analytical usually leads to further delays. The machine at this point represents months of conceptual and formal engineering effort. This effort is first expressed through two-dimensional assembly and detail drawings. The actual fabrication and assembly of the system are done by people who may misinterpret the design intent or specification or may not completely understand the function of a specific station. *It is unwise to assume without evidence that the design is at fault.* It may simply not be executed properly either in fabrication, in assembly, or in sequential control timing. It may be that simple adjustments to the relative position of the station to the fixture or escapement may clear up the problem. The adjustments to position, sequence, or timing of any station during debugging should be done only after analysis and only after ensuring that the original position or timing can be reestablished if necessary.

Nothing is more frustrating than the attitude of "let's move it and see what happens," particularly if there is no record of the original setting. In stations with multiple adjustment features, the tendency to adjust the position or timing of one device after another in a blind search for improvement, without returning the other elements to their original settings, adds to the confusion.

When examining an inefficient station, if we assume the competence of the designer, we should also start with the corollary assumption that the design may not have been properly executed in fabrication or assembly. The first question that should be asked is: "Is this station built in accordance with the design?"

If continued study and adjustment of a station indicates a weak or faulty design, one must avoid pasting bandages over terminal cancer. If every possible attempt is made to ensure that the original design could succeed, and the station still does not work properly, it may be necessary to scrap the first design and redesign the station in a satisfactory manner. This decision will be a real test of the maturity of the customer and the builder.

It is hard for any designer to accept the necessity of admitting his or her creation does not work as planned. Constructive criticism can be ill received. The builder is anxious to complete the project and receive payment. The builder does not want to accept further delays and additional costs. The manufacturing engineering team responsible for the procurement is under increasingly severe pressure from their own management to bring the machine to the production floor. Under these often tense and emotional situations, it is easy to degenerate from debugging efforts to the allocation of blame.

In an attempt to salvage the first investment in the improperly designed tooling, air lines, springs, and thin blades may be added. Extra support plates may be fastened. The essential question is whether or not the station is fundamentally correct and requires additional strengthening or whether the basic design is inherently faulty and needs fundamental redesign. The answer will usu-

ally lie in whether or not the added "fixes" will be durable during the production life of the machine. Will little wires, air jets, springs, or blades stand up to millions of cycles? If a larger cylinder has to be mounted, will the station be strong enough for the additional loads imposed, or will fatigue cause early failure?

There is an additional question that must enter into the decision. Is there truly a better way? *Different is not necessarily better.* Often the decision needs new faces when those who have been long involved for months seem unable to take a fresh objective look at the problem, being mentally conditioned to seeing it from a specific viewpoint. In the author's experience, *customers new to the problems of mechanized assembly tend to give up on a station design too quickly, and builders give up too slowly.* Sometimes the cheapest and fastest solution is a totally new station. The burden of this expense will and should usually fall to the builder, unless there are very unusual circumstances, i.e., customer insistence on a specific station design overriding a builder's objections.

The debugging process is often done under the duress of skepticism on the part of the customer's management. A major psychological problem in the introduction of assembly machines is created during debugging by negative or skeptical attitudes on the part of management. The solution lies in management's understanding of the debugging phase of machine use. In trying to debug and ramp up an assembly machine, one literally starts out by attempting to pull himself or herself up by one's own bootstraps. The first plateaus of productivity are reached relatively easily. Each subsequent plateau generally is a lesser gain in productivity and is harder to realize. For example, usually in the beginning and rough debugging, very little attention is paid to hopper inadequacies other than absolute failure. As machine development continues, hopper performances that might be adequate in the off-and-on initial machine operation may become marginal under full production conditions. The hoppers might have been developed using pre-production samples or without full exposure to the possible variations that occur during full scale production. Efforts will be required to bring feeder performance up to par. The same will be true in almost every other area of the machine. (Figure 11-8.)

It is very important, to realizing the full potential of the equipment, not to show disappointment or dissatisfaction as such problem areas become apparent. If the fundamental machine design is sound, these problems, one after another, expose the stepping stones to final acceptable rates of productivity. They should be discussed and considered as advances in debugging, not as setbacks.

You cannot overemphasize the importance of positive attitudes toward the ultimate success of the system—and the expression of these positive attitudes to those charged with operating the equipment.

If the tendency to adjust without analysis can lead to problems, the completely opposite tendency is equally ineffective. One often sees an attempt to make a new machine run better by constantly pushing the run button, waiting for the machine to improve itself. There is some small element of truth in this attitude. New machines tend to be stiff. A multitude of little sharp edges

Fig. 11-8. As debugging progresses, marginal operation, not noticeable in early stages of development, may require additional work.

and irregularities catch at parts being fed. After a few months of operation, the movement of parts begins to polish feeders, tracks, escapements, and fixture nests. Bearings begin to adapt to their mating slides and shafts. A good machine that is "broken in" does run better and is more productive. This, however, reflects the patina of a well-running machine and is not the way to approach debugging. Problems have causes which with time and analysis can be identified and can be corrected.

There is one other potential troublesome area during the debugging phase. For both valid and invalid reasons almost all machines are built with memory or logic circuits, stopping the system only when consecutive failure occurs at any given station. This type of control mode tends to gloss over the problems of any given machine station during debugging and in particular makes it difficult to identify the cause of failure, to determine whether it was due to part problems, fixtures, or station failure.

In earlier assembly machines with hardwired control panels and circuitry, it was difficult or impossible to disable the memory or logic circuits so that the machine could stop at each failure during debugging and allow a decision to be made concerning the specific cause of stoppage. With modern programmable logic controllers commonplace on assembly machines, it requires only a few moments of programming to disable the consecutive failure mode and thus cause the machine to stop whenever a fault is identified. Initial debugging should be done without memory control circuits operational so that station design efficiency can be determined. Memory circuits should only be actuated after reasonable levels of production have been achieved.

The degree and extent of the debugging effort will be determined not only by operational requirements but by sample availability. The logistical problems of obtaining several thousand sets of actual production sample components for debugging and acceptance runs are most difficult, even more so in concurrent engineering projects. Depending on the cost of the samples all debugging and acceptance runs often have to be done with a few thousand sets of parts. If the machine itself is cycling 20, 30, or 40 cycles per minute, sample availability for all debugging and acceptance may be limited to a few hours of actual machine operations. Very short test runs followed by significant objective analysis are essential. (Figure 11-9.)

During debugging, the *emphasis is on station reliability or efficiency*. The focus during debugging is on the ability of the machine to do its intended functions while accommodating the broadest possible range of component part quality. Debugging should bring station efficiency to the point where an acceptance run for approval to ship the assembly system is practical.

Once all of the major station problems are overcome, the attention should be on achieving full automatic operation to a production level where an acceptance run is feasible.

Using Debugging for Operator Training

The last stages of debugging are ideal times to bring operating people into the project. This is particularly true when the system has operator positions for direct loading of parts into fixture nests, conveyors, or operator-assist devices. Ideally, if operators can spend time at the builder's plant to learn not only the function of the equipment but also how jams are cleared during the last few days of debugging, they will become adept at operating the machine for an acceptance run.

There is often great reluctance at bringing operating people from their home plant to a builder's plant for such training. During debugging they may seem to spend a great deal of idle time while engineers and technicians continue the debugging operations. Such participation, however, gives the future operating team a feeling of involvement and identification that may greatly reduce the learning curve for the equipment on the customer's production floor.

THE ACCEPTANCE RUN

The purpose of the acceptance run seems quite clear: a determination that the machine is capable of sustained production runs and is ready to ship to the operating facility. Often the purchase contract has specified, in most precise terms, the nature of the acceptance run. In the interim period, however, *events tend to weaken the ability of the customer to insist on the con-*

Fig. 11-9. This large shipment of samples is only sufficient for 220 minutes of actual assembly machine operation. Limitation on samples for debugging requires very short runs for analysis or acceptance.

tractually specified acceptance runs. Production samples are often in short supply. The purchasing team seems to be involved in a multitude of projects. Debugging seems to have consumed an inordinate number of the available samples. All of this occurs under significant management pressure to get the machine onto the production floor.

In this confusing context, the customer team may often be faced with the responsibility for accepting a major capital expenditure without a working background as to what really is acceptable. What really is the purpose of the acceptance run? How much can be accomplished on the builder's floor? How much is best left for post-installation acceptance runs on the customer's floor?

The Purpose of the Acceptance Run

The purpose of the builder's acceptance run is NOT to determine that the machine is fully developed. There should be but two acceptance runs before final acceptance: the run on the builder's floor and the run on the customer's floor. *The builder's acceptance run is meant to demonstrate that the machine has reached a point of development where economic considerations indicate that more can be achieved toward final acceptance on the customer's floor than can be realized on the builder's floor.*

Among the more significant factors are the cost and the availability of sample parts, and the potential sales value or the usefulness of the products coming off the assembly machine. Once an assembly machine is capable of operating at a level of 50% or 60% of designed gross production, the consumption of component parts and the cost or value of the assembled product must be of increasing concern to the customer. Each incremental step in improved system productivity, however, becomes more difficult and time consuming, and sample consumption rates increase rapidly. A decision has to be made whether or not the output of the machine can be salvaged or sold to help initiate the return on the machine investment.

The customer, on the other hand, must look for certain levels of capability to justify approving the machine for shipment.

Acceptance Criteria. Before examining the criteria for preliminary acceptance it is necessary to determine whether the equipment is to be shipped directly to the builder's operating plant for direct introduction on the production line or whether it is to be shipped to a customer's machine development laboratory prior to final shipment to the production facility.

The concept of machine development laboratories is not a new one, but is rarely used in a world of shortened time-to-market. The rationale for their existence can be one or more of the following reasons with varying degrees of validity.

- Strong union contracts make it difficult or impossible for the builder and customer engineers to continue the debugging and acceptance procedure of a new system once it is placed on the production shop floor.
- Lack of engineering or toolroom facilities at the operating site, particularly in new plants in remote areas, make continued development impractical at the operating site.
- Proprietary or confidential reasons indicate the need for adding tooling or doing final development in a restricted area other than the builder's floor.
- Continuous expenditure on special machinery justifies the customer's creation of a permanent machine development team to bring machinery up to the highest possible production level in a closed area where quality of production can be closely monitored.

When the machine is to go directly to a customer's development laboratories, it may be expected that acceptance criteria will probably be on a lower level of productivity than when the equipment is to go directly to the final production floor.

Additionally, before going into the specific areas of acceptance, machine construction, station performance, statistical tools, acceptance levels, and spare part requirements, it is well to consider the atmosphere in which acceptance runs are conducted, so that specific guidelines are agreed to beforehand.

Essentially the criteria is for the customer to view a production run of specified duration at a reasonable level of net production. Most builders will agree that the mental perception of an inexperienced buying team on such visits includes expected levels of productivity rarely achieved. *The builder is inclined to view the acceptance run for permission to ship the machine as a continuation of the debugging activity.* Often an acceptance run is stopped in midstream by a builder as he or she recognizes the source of some continuing problem and wants to rectify it immediately. Most customers do not like these interruptions. These opposed viewpoints often lead to significant ill feeling, particularly when the purchasing team is large and their time valuable elsewhere.

The real purpose of any builder's floor acceptance run should be to determine the present level of productivity (Net Production Rate) over controlled periods without further adjustment or correction. *For the first time the two factors of system and station efficiency and average downtime created by each incidence of failure are viewed as an entity.* Short runs, an hour or two at a time, should be scheduled, followed by an analysis of that production run and decision on what corrective action is indicated. For this reason, any significant acceptance run to determine if the sys-

tem is ready for shipment should start with an advance team, followed by the whole group or team. (Figure 11-10.)

Machine Construction. In the Niland book referred to in past chapters, and in the experience of many users, problems in acquisition of special machinery often center on the poor performance of the basic machine chassis and its control system. *Concentration on the tooling and the product may be to the detriment of concern about the underlying machine system.*

Many experienced buyers insist that the first stage of any builder's acceptance run consist of a 24 or 48 hour continuous machine run without any component parts being placed in the machine. This run should determine the condition of bearings, controls, slides, solenoid, cylinders, and lubrication systems. The usual stipulation is that this run should be done without any breakdown or failure and without adjustment. This is a most valid request. A machine unable to complete this test is not suitable for shipment.

The customer team during this early run should particularly look for potential maintenance problems such as fraying or rubbing flexible cords, air

Fig. 11-10. Acceptance runs are costly. Preplanning is essential to ensure that every aspect is reviewed with the limited number of samples available.

lines, and lubrication lines, and the loosening of screws and small pneumatic and hydraulic fittings. The sympathetic vibration induced from part feeders is a special concern.

The efficiency of the machine chassis and its control system must be superb. Nothing else will do.

Station Performance. The high efficiency demanded of the machine chassis and station actuators and its control system are solely the responsibility of the builder. Only the highest levels of efficiency are acceptable. Once station efficiency is considered, we now become concerned about the coupling of the machine to the product. In the acceptance run one looks for the ability of the system to handle the variations in the component parts of the assembly. In the event of station failure for any cause, a prime concern is the time required for the operator to safely restore the system to efficient operation.

It is necessary in this evaluation to determine if the problem is station failure or part quality related. A systems-related failure is completely the builder's responsibility, while responsibility for part-related failures must be determined jointly. For this reason alone acceptance runs should be done in relatively brief periods followed by a complete analysis of that specific period.

Statistical Tools. It is commonplace when accepting metal cutting machines for the customer's teams to do a quality evaluation of the machine's capabilities by a detailed examination of a few parts produced on each fixture. Often the acceptance criteria will be the machine's ability to hold some specific percentage of the parts print tolerance. Experience has shown that such evaluation is a good indicator of future machine performance. Production rates based on established speeds, feeds, and machinability ratings are not usually a cause for concern.

When evaluating the performances of an assembly system in an acceptance run, the study must be oriented along two lines: net production and quality levels of the product produced. Quality evaluation will be complex since it will be necessary to determine whether assembly quality is component part related or caused by

the assembly process. Net production will, however, be easier to document. If the machine is equipped with counters on each inspection function, the trends will be quite obvious.

In an acceptance run, the incidence or rate of failure is only one aspect. In evaluating the machine's productive capacity, each hour's run should be monitored and the following items recorded:

- Duration of the run.
- Achieved net production.
- For each failure, the station, duration of downtime, whether the failure was machine or part related, and lastly whether the failure would have cleared automatically at the next cycle of the machine.
- If possible, record when each new batch of parts is fed to the machine.

Each run not only should be as short as possible consistent with achieving a significant data base, but it should be continued, no matter how poor the experience, for the planned time and not aborted unless damage to the machine will occur.

No objective analysis can be made of machine capability without recording the above suggested data. Adjustments during the run vitiate the recorded data.

Acceptance Levels. What levels of productivity should be looked for in an acceptance run? As mentioned earlier, this will depend in part on whether the machine is to go directly into production or whether it is to go into an interim development phase in the customer's machine laboratory. Also, in part, the complexity of the assembly and the state of sample development will have a significant bearing on the degree of net production realized in the preliminary (or builder's) acceptance run.

It should not be expected that the achievements of an assembly system will exceed the productivity levels of other types of manufacturing machinery. In the Preface of this book it was pointed out that mechanized assembly is rarely "automation" and all that automation implies. It may well be that a given machine will even exceed at time of acceptance the quoted gross rate of the machine; but, in terms of the designated gross cyclic rate at time of acceptance net production of

60–75% will probably indicate a reasonable range for machine shipment. Where parts for tryout are scarce, or where production parts are unavailable and debugging done with prototype parts, 50–60% net production may be sufficient for preliminary acceptance and authorization for shipment.

Terms of payment will have a bearing on acceptance levels. On machines without significant progress payments, it is to be expected that the builder will push for rapid acceptance for shipment. When milestone payments have been made, the customer may be in a position to demand relatively higher net production before granting preliminary acceptance.

Spare Part Requirements. The acceptance run is an ideal time to develop lists of potentially useful spare parts to support the original equipment purchase. During the acceptance run, the fragility or perishability of tooling items and machine components will become evident. The builder's service department can provide useful input in building such a list of parts to be stored in maintenance stocks to back up the machine on the production floor.

Nothing is more disruptive than a major system down for lack of spare parts. To order these parts before the acceptance run may mean that the parts ordered and made concurrent with machine build will be made obsolete during debugging and acceptance run modifications. To delay much beyond preliminary acceptance runs may mean possible long downtime, because of potential damage incurred during machine startup and final acceptance runs. Warranty considerations enter into this decision, but the most rapid routine warranty replacement of a tooling item worth a few dollars may cause production losses of tens of thousands of dollars.

Acceptance for Shipment

Acceptance for shipment should mean that the following conditions exist:

- The machine has reached a point of productivity where there are realizable economic benefits to be gained by placing it in production.
- All stations function at a level where major redesigns or replacement of stations no longer is indicated.

- Modifications which may prove necessary in final debugging and acceptance can be supported by toolroom facilities at the user site or by the proximity or rapid availability of builder-supplied components.
- Both builder and customer agree that progress toward final debugging and formal acceptance will progress more rapidly at the user's facility.

If *all of these conditions are not present, the machine should not be shipped*. During procurement, no single item is discussed more than the details of rigid acceptance criteria. In practice, however, production demands and overall program delays place premiums on delivery, since assembly equipment acceptance often has to wait for samples available only after all fabrication equipment is received.

Machines are often accepted, in deference to customer's management pressure, before they reach acceptable levels of productivity. Premature shipment of a machine will usually not only lead to prolonged delays in placing the machine in production, but will probably mean that the machine will never reach its full potential. In any event, it will command time and expense at the user's plant site, which will ultimately be more costly than the cost of late delivery of a machine ready for operation.

MACHINE INSTALLATION

Somewhere during the shipment of a machine tool, legal title passes to the customer. If the FOB point is the builder's dock, title passes as soon as the machine is placed on the delivery truck. Because of possible shipment damage, some customers insist on and will pay a premium to have the FOB point be their receiving platform.

In any case legal title will pass to the customer upon shipment of the machine but moral responsibility and contractual obligations should remain binding on the builder to complete the debugging process and participate in a final acceptance run on the customer's floor.

Obviously, warranty responsibility resides with the builder and legally so even if there is no definition of this responsibility in the purchase order and its acceptance letter. The necessity, however, to have recourse to the legal terms of warranty means that there is a lack of proper relations during this critical transitional period.

With the shipment of the machine, the builder has completed many of his or her responsibilities, but there remains that of final debugging, completion of the acceptance run, and operator training. These contract responsibilities remain with the builder at a time when the builder's authority and ability to make changes is rapidly waning.

Defining Responsibility

Once the machine has left the builder's plant, equity in the builder-buyer relationship begins to fade. This means that the clear-cut lines of communication between the builder and the purchasing manufacturing engineers now are opened up to include those customer personnel with responsibility for production and plant maintenance. Essentially, the builder's role becomes that of a teacher, using his or her tools of education and persuasion. In fact, this role hopefully began when maintenance and operator personnel were sent to the builder's plant for training during the final stages of debugging and the preliminary acceptance run. Not enough good can be said of the participating role of having operator and maintenance people work with the builder during the last stages of the machine development.

The responsibility of the builder in installation lies in three areas: restoring the machine to the condition in which it ran prior to shipment; continuing the adaptation of the machine to the actual production sample parts (a continuation of the debugging); and training of operational people in routine operating and maintenance procedures. This last area is paramount to obtaining maximum productivity.

Maintenance and Operator Training

The importance of operator training cannot be overemphasized. It is a problem which, if not solved initially, will later raise its ugly head. The general evidence of the problem is the user who becomes dissatisfied several months after a machine is put into production. Production goals are not being met; supplementary hand lines must be maintained, and financial management is up in arms. The irate user wants the builder to drop everything else and send servicemen, engineers, managers—anybody—to fix "a machine that just doesn't work."

Sensible builders would never ship "a machine that just doesn't work." Enough production is usually run in the builder's plant to assure both builder and user that the machine truly is production-worthy at the time of shipment. It *usually turns out that a general complaint about overall machine productivity, rather than specific complaints, is typically evidence of failure to properly train operators on what they should and should not do.* For example, some people cannot control the urge to play with adjustments to see if the machine can be made to work better. It is not at all uncommon to visit a customer after a few weeks of operation, and find every conceivable screw, nut, knob, turnbuckle, and compliant link completely out of its normal position. When something goes wrong, instead of analyzing the problem, someone immediately turns a screw or adjusts a locknut. No one ever remembers where they were originally set and things continue to go from bad to worse. All that is required to solve the problem is to return the machine to its original state.

The "over analysis" syndrome stems from an inability of machine operators, machine repairmen, or operating foremen to accept the limitations inherent in any unique special-purpose machine. There are limits to the tolerances that can be maintained in production on component parts. There are limits to the amount of machine development that is practical before turning a machine over for production use. What is to be expected from such equipment is that the initial net production realized is within a range that permits reasonable justification for the purchase of the equipment. But some users appear to expect and aim for perfect performance. Every time the machine stops, three or four people will stand around for untold minutes discussing the situation, with no one attempting to return the machine to operation.

Successful assembly machines are always self-inspecting. They inspect for the proper actions of their own elements, and for the presence and position of the component parts of the assembly being processed. Upon detecting some malfunction such machines either shut themselves down or switch to an alternate control mode. In any reasonably designed machine, there will be a few random self-clearing failures. But most failures to assemble a product indicate an impairment in machine action or jams caused by foreign material or improperly toleranced parts in the part feeders. Regardless of the control mode, someone must correct these deficiencies and return the machine to operation.

There comes a point where continued analysis of the cause of such failures is counterproductive. Users must accept that such problems occur regularly. They must judge when the machine has reached a mature state of development and then turn from debugging to running the machine and attempting to obtain the maximum production results.

The solution is a simple change in outlook. In many instances, assembly machine servicemen have completely cured production problems by offering operators a cash gift for reaching a certain level of production. To achieve this, the operators were told to stop diagnosing the problems of that machine and attempt to operate it to meet a specific production goal. In almost every instance, this change in orientation and motivation will result in immediate production increases.

Initial training of operators and maintenance people would have forestalled the need for such a "service call." It must be explained that upon turning over a machine to production, *we have switched the emphasis from improving efficiency to reducing downtime.* It must be a management decision that there may be greater economic return in reducing the length of the average stoppage of the assembly machine than there is further analysis of station faults.

This is a difficult thing for operators to understand. During the acceptance runs the builder and manufacturing engineers have attempted to reduce the incidence of machine stoppage by analysis of each problem as it occurs, often with the machine standing idle. Operating people observe this and fall into the trap of believing this behavior is normal on the machine. It must be explained that this process has reached a point where production is more critical than continued analysis.

Great increases in productivity for most as-

semblies (but not matched assemblies) can then be made if the operators are trained to have subassemblies of known quality staged around the machine which can be substituted for defective subassemblies in a few seconds. After the machine is restored to automatic operation, the operator can then attempt to correct the deficiency by hand or discard the defective subassembly. This procedure will work in almost every instance except for serially coded parts such as lock sets or for matched assemblies. The average downtime due to failure to insert parts can be reduced to seconds.

The second significant way that operators can reduce downtime is by being taught to aggressively look for jams in part feeders and attempt to clear them before tracks empty out. Even when it is necessary to stop the machine in order to prevent feeder tracks from emptying, downtime can be significantly reduced.

These two relatively simple techniques can amount to several thousand additional assemblies per day.

Debugging (i.e., continuous improvement) should always continue throughout the life of the machine, but not at the expense of easily attainable required production.

TURNING THE MACHINE OVER TO PRODUCTION

At some point in time the machine must become a production tool, its operation and maintenance a routine procedure. Continued improvements in productivity can come from several sources among the routine maintenance procedures and production logs.

Once assembly machines are turned over to production there seems to be recurring periods of high and low productivity. Production requirements focus management attention on the machine and necessary disciplines are enforced until desirable levels of production are restored. At this point attention turns to other problem areas. The machine production rates continue on for a while and then slowly begin to descend until they reach the point of renewed management attention. At this point the question is asked, what happened? Unless objective data are available, there is no ready response. Production logs are vital to corrective action. Often they will pinpoint the problem as being external to the machine—perhaps personnel changes, but more probably changes in vendors, tooling, or fabrication processes. Modern assembly system controllers can show trend analysis, downtime related to specific fixtures and other data for full analysis.

If a machine production level changes abruptly, the first consideration for improvement in the absence of known damage should be to look for changes in personnel, material, or processes exterior to the machine.

Routine Maintenance

Systematic maintenance procedures are meant to prevent extensive and time consuming repairs. This is a topic in itself and not one to be dwelt on here. There are three areas that should be checked routinely, however, and, if practical, logged on a daily and weekly basis: lubrication system performance, inspection for rubbing flexible hose lines and loosening of their connectors and ferrules, and checking for the tight condition of threaded fasteners. Daily attendance to these inspections will usually prevent extensive machine damage and additionally reveal areas of wear that will require planned repairs at some future date.

Production Logs

Nothing is more essential to the continued performance of special machinery than daily logs of the incidence of failure at each station of the assembly machine. These are vital not only for determining where immediate attention is required, but also, by comparison with previous shifts, for revealing trends that require corrective action.

The scope of these production logs is detailed in the next chapter.

SUMMARY

Successful assembly machines result when the talents of builder and user are combined with each contributing from the areas of their competency. Machine acceptance is done in a series of small steps, each one of which narrows the available choices in subsequent decisions. No decision is irrevocable in itself but time and limited funds make backtracking costly and difficult. Unless the customer participates on a regular continuing basis, they will find themselves carried along on a tide of decisions instead of participating in the decisions which will ultimately determine the actual day-to-day production of the assembly system.

Chapter 12
The Production Audit

INTRODUCTION

One common fallacious assumption, and a source of much past implicit or explicit dissatisfaction with automatic assembly, has been the assumption that assembly machines at time of delivery are at their fullest potential for production. It should be obvious, but is often overlooked, that the actual operational time of any machine during debugging is severely limited. At best the machine has had all major design problems solved and is capable of meeting modest production goals. The best is, hopefully, yet to come.

In order to realize the full potential of any assembly system, it is necessary to continuously observe and objectively analyze the performance of the machines from an operational, maintenance, and fiscal standpoint *after the machine has been placed in production*. Such systematic analysis is thwarted by a lack of management patience and the operational tendency to modify or adjust the machine without objective data of performance prior and subsequent to the modifications or adjustment. This situation is compounded if no records are kept of the original settings or conditions and hence it becomes difficult to restore the machine to the original condition if any suggested modification or adjustment proves unsatisfactory.

A realization that *few assembly systems are ever static in design or performance* is beneficial. Continuous improvement is the only game in town. Logistical and cost considerations place severe limitations on the extent of debugging. Debugging is mostly done in an artificially controlled environment.

Good systems should continuously grow in reliability as one after another minor problems are identified and corrected. This chapter outlines systematic approaches to such system refinement but also outlines the means to rapidly identify any unexpected system deterioration before it becomes critical.

In this book, there has been a continuous attempt to avoid comparison of American efforts toward automation with those of Japan, since there is a totally different operating environment in Japan than that in America and in European countries. It would seem appropriate to record the comment of a senior Japanese manager regarding a speedometer assembly line built for them in the United States. He said, "I show this equipment to all of our visitors and tell them what wonderful machinery it is, for we have never run out of the ability to make it run better." This comment is worth deep consideration, for it recognized the essential point of this chapter, namely, that *any assembly machine when shipped has had a limited (and artificial) exposure to production-type operation during debugging.* As mentioned in the previous chapter, the logistical and financial limits on debugging in the builder's plant often confines initial debugging to a determination that there are no major design errors in the system or the individual stations. Beyond that point, any further debugging achieved on the builder's floor remains worthwhile since any additional progress can be achieved more rapidly there because of the technical and human resource availability than on the customer's production floor.

The second lesson from the Japanese manager's comment is an implicit recognition on his part that upon delivery *his engineering department expected to gradually absorb the continuing responsibility of ensuring that this major investment provided the maximum possible return on investment.* It meant that his company fully understood that they had to assume the continuing analytical task of determining the causes of lost production and making the minor but significant modifications to parts, part storage, fixtures, and working stations, as well as machine sensors, to reduce these losses.

This analytical observation of machines is a most difficult job. There is a tendency when observing newly installed machines to want, upon identification of a problem, to do something physical immediately, without any objective analytical assessment of the present status or determination of the problems for resolution on the basis of a priority established by the severity of each type of problem. It should be noted that it is not practical to assess problems based on one or two hours of operations, for the severity and location of problems with assembly machines often changes from day to day. (Figure 12-1.)

Customers should strive to gain objective data and analyze this data before modifying the machine immediately after acceptance. Various systems performance reports are useful analytical tools.

SYSTEMS PERFORMANCE REPORTS

Analysis needs objective data, but such data must include sufficient information if it is to be

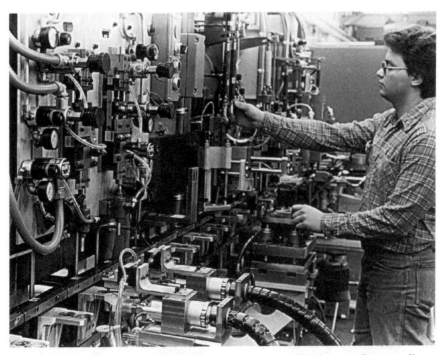

Fig. 12-1. Analytic observation will prove more fruitful than endless readjustment just to see if it will run better.

used properly. Downtime length without noting the cause is useless. The number of failures to insert any given part is worthless without recording its impact on production (i.e., average downtime × incidence of failures). This topic has been covered in Chapter 4 and in the previous chapter.

The Necessity of Objective Data

One of the most difficult challenges most assembly machine builders must face up to when they receive post-installation complaints about poor net production is trying to obtain sufficient data to analyze the problem.

Net production is acceptable or unacceptable only when stated relative to gross production capability, and that is tied directly to the time period in which there was an attempt to operate the machine. A machine operating at a gross rate of 40 assemblies per minute is theoretically capable of 19,200 assemblies in an eight-hour shift. This is true, however, only if the machine is run for eight hours. If the machine produced 16,000 assemblies in that period, it is running at 83% net production if the attempt to run the machine was for eight hours. On the other hand, if it was run only 415 minutes of that eight-hour shift, this same net production of 16,000 is 96% of possible gross production. A determination of how many minutes occur in a shift in which there is an attempt to run the machine is perhaps the most difficult piece of information for any builder or customer to obtain.

Energizing a START button on the assembly system will turn on ancillary equipment such as feeders and may start a time clock running, but is no real measure of attempts to run the machine. Often the start button circuit is left engaged during work breaks, discussions, and other activities without any attempts to engage the RUN button. Measurement of actual time in which there is an attempt to run the machine is an essential element of the required analytical data and will depend on continuous monitoring by trained observers.

A second element often difficult to measure is what portion of idle time on a machine is due to poor scheduling, unavailability of part pieces, and failure of services such as compressed air, electrical service, etc., or cooling water, and what

portion is due to mechanical breakdown of the machine system.

Many readers will balk at the preceding paragraphs, but most assembly system builder's experience has shown that repeated management complaints about assembly machine productivity usually find their basis in a failure to recognize that lack of labor, parts, or services may in itself be the fundamental problem with the machine's apparent lack of productivity.

If labor, parts, and services are available and there is no electrical or mechanical problem with the machine, failure to attempt to run the machine might lie in the lack of motivation or education on the part of the assigned machine operators and their first line supervisor. It is hoped that management has informed workers about the assembly system's potential role in maintaining the profitability and viability of the company. Failure to advise workers and line supervisors about the purpose and function of the machine is ill advised. Participation, where possible, of direct workers, maintenance and department supervision at the builder's plant during the debugging phase will generally be amply rewarded upon installation by shortened learning curves and by their attempt to demonstrate to others in the plant what the operators have learned in their visit to the builder.

Incentive pay and creation of special job descriptions for machine attendants and maintenance personnel can be of vital assistance in maintaining motivation. If there is a hesitancy to upgrade the automatic machine operator's status, one might question what salary levels would be commanded by the supervisors and operators of a data processing computer of equivalent purchase price.

Motivation will require management skills, but education of the operator can usually be done on a lower level, usually front line supervision.

Operators and attendants *must accept the fact that failure to feed or insert is not unusual on any one-of-a-kind assembly system and will probably continue to occur (hopefully with lesser frequency) indefinitely*. Their function, the job for which they are paid, is not to stand there and watch the machine run. They are paid to ensure proper levels of parts in the feeders, to remove defective parts and foreign material from feeders and transfer tracks before jams cause assembly failure and stoppage, and most importantly, they are paid to restore the machine to operation as soon as possible in the event that jams or failure to insert or join parts properly shuts the machine down.

It is not possible for the assembly system to monitor effectively all of the reasons for downtime such as poor scheduling, part shortages, lack of adequate services, or poor motivation. Necessary input can only be obtained by monitoring of operator performance by those skilled in industrial and human engineering techniques. It is possible, however, to precisely monitor that loss in productivity from machine stoppages and determine whether it is created by lack of motivation, lack of education, or poor machine design or refinement.

Once one can determine through observation and study why the machine is not being run and apply whatever corrective action company labor policies and or contracts will allow, one then can proceed to assess machine related stoppages.

The data required include the following:

• Total number of attempted machine cycles.
• The number of completed parts.

• The number of failures detected by each inspection station, whether it be part insertion, joining, functionality, or completed ejection that is being monitored.

This data is easily acquired through conventional relay-actuated counters; where programmable controllers are used to run the machine, these counts may be acquired quite simply from the CPU either by electrically operated digital counters, by command actuated displays on programming units, or by data highways to some terminal or printer. (Figures 12-2 and 12-3.)

A word of caution, when observing failure counts. If a machine is programmed to stop immediately after detection of some assembly failure, it can be assumed that the counts are indicative of actual failure patterns. That is, a single failure will be recorded as a single count. When the machine is programmed to stop after a specific number of failures, there is a strong possibility that one single incident (i.e., a jam in a feeder track) will cause several failures to show on the counters. One jam will cause several bad assemblies. This, when improperly programmed, may indicate a statistically incorrect number of incidents of failure.

Fig. 12-2. A fault count display panel. These resettable counters show the number of failures detected at each inspection station, but do not show downtime.

Fig. 12-3. Sophisticated programmable controllers record data for production control and machine analysis. This unit monitors all of the subsystems in a three machine line and provides hard printouts of machine performance.

One can also note how many incomplete assemblies occur, in machines with memory-type control systems, on second and third attempts to feed after the first failure of any given work station. It is very difficult to make any attempt at diagnosis on a machine using a memory-type control system with downstream lockout features. In debugging and analysis, stoppage after a single failure will provide the most reliable data. Only when enough data is available to serve as a control, can one go to the memory mode of operation to determine if it offers better net production.

Frequency of failure at each individual station and its effect on net production in the overall system as shown in total cycle and net good part counters, however, are only half the data required for objective analysis. Frequency of failure is not overly significant without knowledge of resulting downtime. Twenty insertion failures an hour, which produce downtimes averaging 13 seconds, will still leave a machine with 93% of its production capability. If the average downtime goes to 23 seconds, 87% of the capacity remains. If the average downtime is 60 seconds, only 66% of the capacity remains.

Avoiding the Adjustment Syndrome

It is not at all uncommon for a builder to receive a complaint, several weeks after machine acceptance on the customer's floor, about a sharp drop in net production since the service engineers who installed the machine had returned to their home base. Almost without exception this pattern can be attributed to one of two causes. New operating people have been assigned to the machine without training, or secondly maintenance people have made adjustments without recording original positions of controls, turnbuckles, calibrating screws, and the like. When these adjustments do not prove to be the sought-for cure, the adjusted elements are rarely returned to the original state.

It should be noted that few builder's service engineers are thrilled about being away from home. They may, in order to keep installation visits as short as possible, neglect their primary role of education and training in order to gain rapid acceptance of the equipment by doing actions normally done by operators and customer maintenance people who are being trained. When this happens, the systems production levels will fall apart within days of the departure of the service engineer.

If an assembly has been running well and suddenly a good station becomes a problem, the very first determination should be to look for some change in the component parts. Different part vendors, different molds, different materials, different humidity or temperature may all prove to be the culprit. The second place to look is for possible reassignment of machine operators or technicians. Only after that should one look to the machine itself. Look for loosened lock nuts, broken fingers, etc. Only after this is done and nothing found should any adjustments be made to the station. Such adjustments should be done only after marking, in some way, the original position of the adjusted component.

Performance Monitoring Tools

The inspection devices found in well-designed assembly machines provide most of the data input required for analysis and improvement of assembly machines in production. It is the extraction of that data in usable form that is important. Frequency of station failure occurrence can be read from counters or more sophisticated display units or printers merely by counting the changes in open or closed condition of the output terminals on the machine's inspection devices.

Increased frequency of failure to insert a given component when compared to normal failure rates may be the first sign of component change, station wear, or misalignment.

Recording of each inspection counter in a way that provides for ready comparison with established or accepted norms may pinpoint areas requiring examination at the first convenient period. Log sheet setup should facilitate this comparison and all logs kept chronologically. Changes in parts vendors, molds, or molding press operating forces should be duly noted on these sheets.

It is worth repeating, at the risk of boring the reader, that the frequency of failure to insert parts must be evaluated as insignificant, annoying, or absolutely harmful solely in *the light of the downtime such failures create*. Counters alone are not sufficient. Timers or "clocks" should monitor the seconds that elapse from the instant the machine stops until it is restored to automatic operation. Since it can be assumed that all new machines will use programmable logic controllers for over-

all control purposes, monitoring of insertion failure and downtime caused by such failure becomes quite easy. Most CPU's have both internal timers as well as counters and can readily time many separate events to tenths of a second. Printouts of elapsed downtime as well as insertion failure frequency will quickly establish a priority to the examination of those areas where machine productivity can be improved.

One high-volume consumer product producer recently reprogrammed his programmable controller to include acceptable downtime periods for restoring a machine to operation after detection of failure to insert or join components. The programmable controller then records all stoppages where restoration time exceeds established norms for machine operators. If these become frequent, the operator is instructed in better procedures for restoring the machine to operation. If such education does not result in reduced downtime, the operator is reassigned. Such programming makes a machine capable of monitoring human performance. Such performance norms cannot be reasonably established until some production experience is gained. Assigning restoration goals to operators leaves the engineer with the task of reducing these goals or targets for systems restoration by small improvements to the machine.

Some experienced customers have identified that failure to insert is very often related to defective or worn fixtures, rather than to parts problems or transfer device failure. They insist that any insertia failure be identified by frequency, downtime, restoration time against established norms and frequency of failure against specific workholding fixtures.

Counters, timers, and logs are the basis for three types of audits: downtime, quality, and fiscal. Each examines the same problems from different viewpoints, but one leads into another. It is difficult to audit the quality performance until there is some certainty that the machine has begun to live up to its performance potential. Fiscal audits assume the production of a salable, quality product.

DOWNTIME AUDITS

Once the weak links in a system are identified by performance logs or audits, we must work on two problems: reducing frequency of failure and reducing average downtime caused by such failure. To address one without the other will not work.

Experienced, analytical visual observation is now required to identify the causes. If downtime varies from operator to operator, education and assignment of those workers with required dexterity and analytical ability may prove to be all that is required. Where downtime is extensive, the solution may lie in improving the operator's visual or manual access. Access ports, access windows, modified guarding, and simple tools all can assist in quickly locating defective parts and removing them from tracks, feeders, and fixtures. In a well-designed machine, particularly where all transfer tooling is withdrawn prior to index, a long downtime will probably come from four major areas:

1. The trapping of defective parts or foreign material halfway down the transfer rail.
2. Emptying out of long feeder transfer rails because a part is wedged between the parts feeder and transfer rail.

3. Poor visual and/or manual access from the operator's normal work position to troublesome stations because of poor system layout.
4. Poor visual or manual access to feeders because of sound guarding or mechanical barrier guarding without convenient access ports. (Figure 12-4.)

Downtime can usually be reduced through education and better access. As simple a thing as having partially coupled assemblies that are complete up to a specific point of assembly strategically located around the machine can do much to reduce downtime if quality considerations will allow. Matched assemblies cannot use this technique. When the system stops because of an insertion failure, the operator can remove this defective assembly, insert a good subassembly, restore the machine to automatic operation, and then diagnose what caused the assembly failure. A tremendous amount of downtime comes from operators attempting to correct the condition on the defective assembly while it lies in the stopped machine. Correction of defective assemblies must be made, if possible, while the machine is oper-

Fig. 12-4. Access ports allow an operator to clear jams without the need of removing the sound enclosure.

ating. Corrected assemblies can then be placed where they can be used for the next stoppage.

Discovering Weak Designs

If *excessive downtime* can be attributed to defects in operator motivation, education, or access, *excessive frequency of failure* can be attributed to machine design or part quality. Once the machine is built and debugging commences, one usually finds unforeseen problems with parts as a major source of machine stoppage. *Debugging is usually the systematic modification of station design or dimensions to accommodate acceptable variations in component parts as they become evident.*

This debugging (continuous improvement) process never ends. It is a recognition that at the design stage, particularly in a simultaneous engineering environment, it is very difficult to visualize all of the variations that can occur in otherwise functional components of any product and to foresee their dynamic behavior in feeders, transfer devices, and fixtures.

The human eye cannot often detect exactly what is happening to the parts which fail to be mated, but seem perfectly good. Jogging a machine for observation can be of limited value, particularly where part behavior rather than alignment is suspected. High-speed photography may prove useful, but one is forewarned, watching high-speed photography is stupefying. Like the old saying, "sin in haste, repent in leisure," watching high-speed photography at slow speeds is enough to try one's soul.

New electronic replay video cameras for industrial use can be much more helpful particularly if they can be stopped and replayed instantly after failure. This equipment is unfortunately quite costly, but its use may resolve major problems in costly equipment. Such observations may be the only way to identify what is happening in failures occurring at high speed.

Improving Routine Maintenance Procedures

Downtime audits may reveal the beginning of wear, misalignment, improper lubrication, or other problems which, if not caught early, can lead to severe systems damage and prolonged downtime. Induced vibration and the machine

design compromises that occur because of inertia and necessary tool size also may be spotted in downtime audits.

A recent study of several hundred assembly machine maintenance procedures revealed several interesting facts:

- Efficient producers leave room in each day's schedule to correct what is beginning to go wrong.
- Almost no one follows preventive maintenance procedures.
- It is questionable whether preventive maintenance would increase productivity.
- Machine cleanliness per se seems to help, but does not significantly reduce maintenance costs.
- Tightening of ferrules on air lines and lubrication lines and checking fasteners for tightness seems to be the most useful maintenance procedure.

QUALITY AUDITS

Downtime audits essentially pertain to the functionality of the assembly machine and its proper operation. Functionality of the product produced on the machine is another subject of post-installation audit.

It is assumed that one major beneficial aspect of automatic assembly is product consistency. As in every other manufacturing process, it is much easier to maintain product consistency if the process is mechanized than when it is done by hand.

Consistent quality does not mean good quality; it means consistent quality. Whether that is good or not is dependent on whether the product falls within the control limits established for that product. But if consistency can be established, and only when it is established, can moving it within the control limits and keeping it there become possible.

There has been incidental mention throughout this book of the debt the American assembly machine industry owes to large-scale assembly mechanization pioneered by the U.S. Army Ordnance Department in the late 1950's and early 1960's. A significant part of that development was recognition of the quality assurance provided by well-designed mechanized assembly systems and the ability to inexpensively monitor product quality and document that quality as a byproduct of automatic assembly. The necessity for statistical sampling and destructive product testing was drastically reduced when the awareness of product consistency due to repetitive machine operations became apparent.

Quality audits are necessary to ensure that production on assembly machines remains in control. Documentation of the presence and position of each component and the functionality of the assembled product becomes increasingly important if the product being assembled is to face any significant liability exposure or high warranty costs. Audits from assembly equipment operation pertaining to product quality provide vital insight into the efficiency of quality control procedures on component fabrication.

Poor efficiency and low net production need not come from machine related problems, but often are due to lack of discipline and broad fluctuations in product component fabrication. Subjective opinions are not helpful here, but objective data properly logged concerning assembly failures due to component quality can only be beneficial.

These objective data can often track field problems blamed on poor assembly practices back to a basic problem in fabrication of components.

FINANCIAL DATA

One widely respected manufacturing manager, upon completion of a sweeping project to mechanize the assembly of every feasible aspect of all major product lines, was asked if he made any major errors or wished he had done something differently; he replied, "If I had to do it over again, I would have drastically scaled down my estimates of return on investment." Shortly thereafter, the Vice President of Finance of one of America's largest and respected manufacturing companies commented publicly, "Whenever any of my divisional comptrollers complain about the results of post-installation audits on manufacturing equipment, I ask them if they have audited the savings realized on their last data processing computer installation."

Missed targets are endemic. Projected savings are often overstated, but so, as well, are many corporate goals for return on investment.

The post-installation financial audit for any assembly system can serve two important purposes. It can act as a spur and a goad to better performance of the assembly machinery. It can also serve to better refine future assembly projects. To serve this latter purpose, it is not enough to audit the system as a whole, but to analyze in a fiscal sense what types of return were generated by various elements of the overall project.

Over-capitalization and consequent over-mechanization are even more deadly toward profit generation than their counterparts, under-capitalization and under-mechanization. The latter course can at least be corrected.

Fiscal audits used as blame assignments are worthless. As reminders of the need for greater performance of installed machinery and as guides to the profitable level of mechanization in future projects, they are extremely valuable.

SUMMARY

This chapter pinpoints the area least understood by many automatic assembly system user: the need for continuing development of the potential of their installed machinery. No degree of experience in automatic machine design nor skill in debugging can alone ever expect to bring a complex assembly machine up to top levels of productivity.

Many customers are willing to accept the level of productivity reached during installation acceptance runs as indicative of maximum performance. Others look at the departure of the builder's installation team as the beginning of their opportunity to refine, improve, and increase the capability of the new machine.

It is an unexpected but common occurrence to find that assembly system productivity improves with little or no machine modification after several months of operation, solely because the machine has begun to wear in and adapt itself to component part variations. This breaking-in, together with increases in operation skills, seems to preclude the absolute necessity of continuing effort toward machine improvement. This functional improvement phenomenon will often continue for some years before wear and tear begin to set in a gradual deterioration of machine productivity. This "wearing in" phenomenon should not be viewed as replacing the need for a performance audit program. Instead it should indicate how *very minor polishing and adaptation can significantly improve machine productivity.*

These various audits require almost the full time of an engineer to be truly successful.

Assignment of a newly graduated engineer to the full time task of analytical audit and subsequent modification of a newly purchased machine can be professionally rewarding to the new employee or intern and financially rewarding to the customer. It will provide to participants an insight into the delicate balance between product design, practical manufacturing tolerances, operator motivation and education, and good tool design practices in a successful manufacturing operation.

Chapter 13
Automated Assembly in the 21st Century

INTRODUCTION

There is little danger in forecasting the trends and technological developments that will determine the face of automated assembly systems in the 21st century. As this book is published, systems are in the early stages of design that will be delivered in 1997 and early 1998. As mentioned earlier in this book, the American assembly system industry is a fully mature industry serving not only North American needs but increasingly supplying the needs of major manufacturers in Europe and Asia as well as emerging countries. This industry, which had its seeds in several developments following World War II, has weathered recessions, negative management reaction to early primitive machines, the rise and fall of the second robotic era, and now finds itself involved at the very center of three major manufacturing management concerns.

The first of these concerns is the determination of major manufacturing corporations to reduce the number of permanent employees both in the blue collar work force and in the middle management ranks, while at the same time upgrading corporate skills in their core competencies. The second major factor is the global drive for product quality in which both quality verification and documentation can often only be achieved through automated manufacturing systems. This drive for quality has forced those companies which have relocated into low labor cost areas of the world to reconsider their attitudes toward major capital investment in automated manufacturing equipment.

The third major management concern is the successful implementation of simultaneous engineering which will require totally new types of relationships between the major manufacturers and their manufacturing equipment and component suppliers.

Some early management responses to these concerns have already produced severe operating problems. Downsizing of work forces, championed in such books as *Reengineering The Corporation*, has led to the emergence of the term "corporate anorexia." Many companies laying off senior engineers and operating managers have been forced to recall these same people from retirement in order to maintain continuity of manufacturing skills based on long application skills and experience.

A few major manufacturers, using their size and market dominance, have attempted to push equipment and component suppliers into partnerships reminiscent of the cotton plantations of the early 1800's. These excesses, and the problems they cause, however, are rapidly apparent and will not affect the overall thrust and direction of the three concerns listed above. This chapter is intended to identify those thrusts and developments which will influence the procurement process for major automated assembly and testing systems designed to be integrated with all of the other operations in a manufacturing enterprise.

WORK FORCE ISSUES

The ever changing role of engineers involved in the tooling and process development in manufacturing and those that implement these processes on the factory floor has led to several significant studies about the functional role of manufacturing engineering in modern manufacturing enterprises. Several aspects of these studies address current problems in the now existing work force, and the future roles of this work force in specifying, procuring and implementing major integrated assembly and testing systems and operating these systems; implementing continuous improvement programs while at the same time responding to rapid changes in the marketplace; and in new product development using all their skills to ensure that purchased machines have inherent flexibility or "agility" and maintain that flexibility or agility during the operating life of such equipment.

Manufacturing Engineering

Manufacturing engineering, as a professional career, emerged from those people in operating factories responsible for the development of the tools and processes for new products often designed elsewhere. These tool engineers had skills which were empirically developed, not only to meet the challenges imposed by new product designs, but the limitations of then existing machinery.

Today's manufacturing engineer is perceived by corporate executive officers and senior manufacturing management in entirely new roles. The Society of Manufacturing Engineers interviewed several thousand senior manufacturing executives to determine what these roles would

be and what educational requirements there would be for the work force of the future. This study, "Count Down To The Future: The Manufacturing Engineer In The 21st Century," was more popularly known as "Profile 21." In the few years since it was published its accuracy of prediction has become widely recognized throughout the global manufacturing community. It states that manufacturing engineers will fill one of three different roles: manufacturing strategists, technical specialists and operations integrators. Each of these roles will require a different balance of knowledge depth and breadth in education and developed skills.

The depth of skills will include the traditional technical engineering disciplines while the breadth skills include non-technical capabilities including communication skills, foreign language fluency, team work and the ability to deal with broader business issues. These emerging roles are shown in Figure 13-1—Multiple Roles of the Manufacturing Engineer. Those people emerging or developing into the roles of operations integrators will need a balance of breadth and depth skills so that they can interact effectively with the technical specialists and the manufacturing strategists. (Figure 13-2.)

Manufacturing strategists will need significant breadth skills focusing on *why* something is to be done or *what* should be done rather than *how* should it be done. They are intended to break down the existing barrier walls between financial and marketing management and those people involved in manufacturing of the product, giving manufacturing a voice in strategic planning.

Technical specialists will need very focused in-depth skills to ever more deeply look into traditional and emerging technical disciplines. There is some belief expressed in this study that the importance of technical specialists will decline as the availability of expert systems becomes more common. In many cases, available software usable on PC's and lower cost work stations is already empowering engineers with such expert systems.

Perhaps the most significant aspect of this division of roles within the manufacturing engineering community will be the significant reduction of

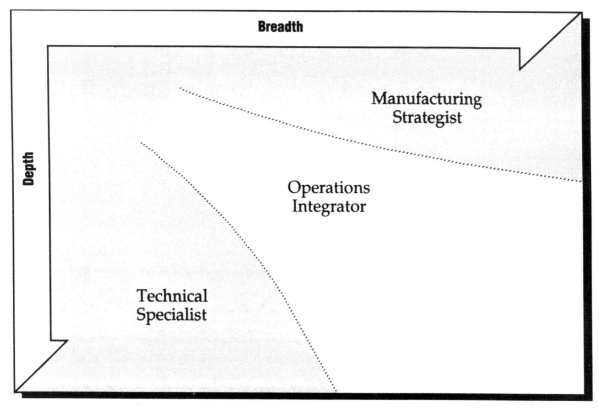

Breadth

Depth

Manufacturing
Strategist

Operations
Integrator

Technical
Specialist

Fig. 13-1. Multiple roles of the future manufacturing engineer.

operations or system integrators finding employment in major corporations. Instead this study suggests, and developments of the last few years prove, that more and more people having integration skills and capabilities in combining and merging the tools and technologies will be employees of consulting firms somewhat akin to architectural engineering companies. Companies more and more are eliminating their in-house process development capability to focus on the core products that are their major source of revenue. They are increasingly willing to hire on a project basis systems or operations integrators for specific capital equipment projects and then release these people when the specific program is completed.

Entirely new forms of contracts will have to be developed to engage system integrator companies and their personnel in a fair and equitable way. New practices will have to be developed so that such firms can be engaged or placed on retainer in the same way that manufacturing companies engage banks, lawyers and accountants. The major players in system integrations work at the present time are not merely consultants, but instead represent a changing role for established assembly systems manufacturers in which their engineering function is no longer a secondary role but a major product for which they will be engaged, while their ability to implement their engineering lessens the risk of new process development.

College Trained Systems Operators

It has become apparent that the sophistication and capital costs of major integrated systems will require ever greater educational levels for systems operators or attendants.

In process industries, such as modern steel mills, it is not at all uncommon to find that those operating the production equipment are, in fact, graduate engineers who act not only as operators but maintenance technicians with an additional

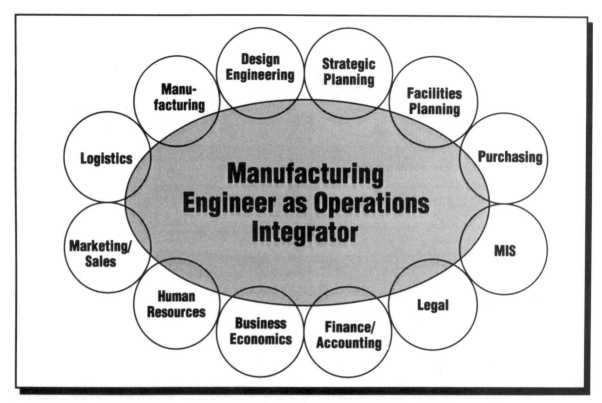

Fig. 13-2. Manufacturing functions to be coordinated by an operations integrator.

responsibility of identifying and implementing continuous improvements on such equipment.

It will not be uncommon to find that assembly systems of the future will be operated by highly skilled and trained people capable not only of routine operation but continuous improvement and maintenance of the system itself.

In those instances where operators or technicians will be required without college level skills, there will still remain the need for intensive training not only on the basic machine system but the sophisticated gauging and testing equipment in specific working stations. The unique labor agreements made between General Motors and the automotive unions at the Saturn plants reflect this growing recognition of the need for dedicated skilled operators protected from bumping or transfer from their assignment to highly sophisticated automated assembly systems. It can be expected that future labor contracts will reflect the need for job permanency in skilled major system operators.

SIMULTANEOUS ENGINEERING

During a major conference on implementing simultaneous engineering conducted at the University of Southern California, speaker after speaker emphasized that implementation of design for manufacturing (DFM) is the beginning or initial phase of simultaneous or concurrent engineering. The second major thrust of the conference was a review of inter-organizational issues raised by the implementation of simultaneous engineering processes by corporations who are rapidly reducing their own internal ability to develop processes and implement those processes. This section addresses the current status of design for manufacturing and the contractual problems of implementing simultaneous engineering by companies who have reduced or eliminated significant portions of their process development capability.

Design for Manufacturing

Designing for assembly (DFA) is now well into its third phase. Phase I involved the redesign of components beyond the needs of product function to facilitate parts feeding. This early phase usually involved redesigning components on existing products to affect assembly cost reduction.

A natural evolution of the interest in designing for assembly was a recognition that most design engineers ignored the problems of product assembly, leaving it to shop floor supervision. A second phase, therefore, of design for assembly focused in on redesigning overall products to facilitate assembly.

A systematic internal study done by General Electric indicated that the vast majority of customer complaints could be attributed to problems which arose (or could have been detected and eliminated) in the assembly process, or where overall assembly design weakness could bear primary responsibility for customer complaints.

The internal General Electric corporate response to designing for assembly represents a unified approach to reducing both assembly costs and assembly related problems in product quality. It involves not only solutions based on both reducing the number of components in any assembly and also reducing planes of access for component insertion, but its own knowledge base. In particular, it emphasizes the need for management participation in each phase of product and process design.

Today we are in a very mature phase of DFA, the integration of design for assembly into design for manufacturing. Activities in this area address the integration of fabrication and assembly into a total work cell configuration. The full realization of increases in efficiency and profitability demands a full participation of assembly design specialists into an overall team effort for product and process development, solution of production problems, packaging and delivery and marketing and customer relationships—in other words, not DFA but DFM.

In order to fully convert DFA into DFM, manufacturing and design management must address six activity areas which forward-looking engineering and production management cannot ignore: (1) designing product components for reduction and material handling costs; (2) designing both products and processes for future flexibility and changing market requirements; (3) where necessary, addressing the specific requirements of matched or selective assemblies; (4) designing products for a real time (or developing) product quality measurement; (5) designing production processes to accept retrofits of emerging technology; and (6) addressing design problems in light of changing roles of manufacturing engineers and corporate organizational structures.

Designing for Future Flexibility. When designing products and processes one must consider that in today's global economic environment and in the light of ever more aggressive attempts to achieve the best time-to-market, one must consider in all tooling procurement and product design, the requirements for future flexibility, both in product and process, and ever changing market requirements.

Designers who ignore their responsibility for complete consultation about realistic product life cycles with their marketing co-workers, who proceed without discussions with their own research departments, and with those in their corporation involved in governmental liaison, are exposing their company to rapid obsolescence of major capital investments. When choosing among a variety of competitive process machinery and individual stations on complex systems, the design team must factor into their equation the future utilization of such tooling and such production machinery and the capability of such equipment for retrofit in an ever changing marketplace.

So much emphasis is placed on time-to-market that most engineers have ignored the lessons demonstrated in Japan, namely that time-to-market must be balanced against return-on-investment. The technical ability to redesign a major product every year must fall to the valid corporate insistence for significant return-on-investment, not merely to recapture the tooling costs for new products, but to derive a major profit from such capital investments.

Designing for Matched or Selective Assembly. An increasing number of design-related engineers and management must face up to the fact of the increasing use of matched assemblies to achieve optimal performance of critical products and components, and the related area of selective assembly, for small batch assembly of differentiated but similar products.

Increasingly, the market demand for product quality and performance means that components of an assembly must be matched to the specific dimensions of other components. Increasingly in high performance products such as fuel injectors, the fabricating capability of present day machinery cannot meet the functional design requirements of special tolerance components of an assembly. It is necessary, therefore, to manufacture a large number of components, classify them into presorted categories of specific tolerances and then match preselected components with one another to achieve the overall functional design tolerances of the assembly. (Figure 13-3.)

Unfortunately, most people involved in product design remain ignorant of the broad body of research captured by the American Society of Mechanical Engineers in the areas of matched and selective assemblies, and in the studies that involve the imbalances between optimal design tolerances and achieved manufacturing tolerances and the more basic theories of tolerance distribution. ASME has a full library of studies by distinguished academic people in this area. Those who ignore this research expose their companies to severely increased costs and delays in process development.

Designing for Quality Measurement. Another area concerning design for manufacturing and assembly today must involve considerations of real time product quality measurement. Before any commitment to fabrication tooling for components, there should be a team effort to determine the design considerations both in component and product design for access for physical contact, visual access or electronic or sonic coupling to both the individual components and the emerging overall product so that the production equipment can determine quality levels at each incremental step of value added operations. This

Fig. 13-3. Premeasured injector components are fed on demand to meet dimensional requirements.

is perhaps the most ignored area in most product designs. (Figure 13-4.)

A second area that is often ignored is to ensure that the proposed tactile and non-tactile sensory or testing equipment on automated assembly machinery can accept major deviations in product assembly quality without damage to the sensors or contamination of the sensors themselves.

Designing for Continuous Improvement. One common problem in the development of large production process equipment involves unrealistic market expectations of emerging technology. In the early 1980's, American assembly machine manufacturers reaching a significant point of maturity in the overall global market had to address the unrealistic market expectations of robotic, vision camera and laser capability and readiness for industrial usage. There is a delicate balance between proceeding with existing and proven technology and wasting overall quality and cost

reduction possibilities by integrating promising but unproven technology into production lines.

One solution to this problem is to ensure that proposed assembly equipment proven in the marketplace has built into its design retrofit capability for emerging technology when such technology reaches that point of maturity for reliable multi-shift operation on the factory floor. Table 13-1 lists a forecast of emerging developments that senior management expects will be implemented rapidly.

Three significant areas of system procurement specifications emphasizing future upgrading would include providing on assembly system design the capability for ever increasing requirements for real time management information system interfacing. Secondly, any equipment purchased today must provide in its control system the ability for real time feedback to fabricating machines from trend analysis quality software in work cell configured systems. Lastly, in determining whether or not to incorporate emerging

Fig. 13-4. A battery of leak testers is dependent on the product design to permit rapid measurement.

TABLE 13-1. FORECAST OF TECHNOLOGIES USED IN MANUFACTURING—SME PROFILE 21

Technologies	Absolute Percent Increase	Percent Currently Required to Use	Percent Required to Use in 2000	Growth Multiple
Expert systems, artificial intelligence and networking	36%	11%	47%	4.3
Automated material handling	35	23	58	2.5
Sensor technology, such as machine vision, adaptive control and voice recognition	35	16	51	3.2
Laser applications including welding/soldering, heat treating and inspection	34	17	51	3.0
Advanced inspection technologies including on-machine inspection and clean room technology	25	32	57	1.8
Flexible manufacturing systems	24	32	56	1.8
Simulation	23	17	40	2.4
Composite materials	20	16	36	2.3
CAD, CAE, CAPP, or CAM	13	56	69	1.2
Manufacturing in space	11	2	13	6.5
Bio-technology	7	1	8	8.0

technology onto a system under construction, or whether or not to delay that until the system is proven capable of operation on the factory floor, one must learn to be conservative and realistic about actual development times for emerging technology.

One example of this might be the current status of industry suitable machine vision systems today and the promises of ten years ago. Millions of dollars of profit have been lost over this period in placing unproved machine vision systems onto production equipment. Today's mature vision technology can be considered user friendly and acceptable for factory floor operations. Unfortunately, much otherwise capable production equipment is hobbled by premature decisions concerning the true production status of emerging technology.

Designing in Reengineered Organizations. The last area of discussion is addressing design problems in the light of changing roles of manufacturing engineers and the very changing roles of

corporate structures, particularly in the United States. In the landmark study done by the Society of Manufacturing Engineers in the late 1980's, commonly known as "Profile 21," it became increasingly clear that more and more manufacturing engineers will be employed on a project basis by consulting companies rather than serve as full time corporate employees.

In this transition, as companies struggle to be globally competitive, much of the resident corporate knowledge base of specific products and processes has been and is being lost. Few engineers in their professional lifetime will work on more than a few major capital investment programs. Because of this loss of corporate knowledge base, developed through experience and handed on from generation to generation, it becomes increasingly important that those involved in any aspect of the design configuration must utilize a variety of check lists to ensure that major aspects have not been ignored or forgotten.

Probably the greatest problem of all for American corporations in any team effort to design for manufacturing is the tremendous emphasis on accounting control of manufacturing processes. Both in Europe and Japan where there are much flatter organizational structures, where senior management usually comes from an engineering background, there is a greater recognition that concurrent or simultaneous engineering means that capital acquisition authority must accompany product and process design decision.

Strangely enough, until the middle 1970's, this was a benchmark of American success in which capital equipment decision responsibility and procurement authority were integral to manufacturing operations. Today America must address itself to the fact that its engineering does not lack design capability nor does it lack implementation capability, but primarily it lacks procurement authority at a time and place when market share might be easily grasped. The end result is that this lack of engineering empowerment places American companies increasingly in the process of recapturing market share, far more difficult than capturing market in the first place.

Supplier/User Interaction

The current and future value for any manufacturing corporation of simultaneous engineering will depend on that corporation's ability to develop effective fair and equitable relationships with their entire supplier base but most importantly with those contract organizations such as consultancies or assembly systems builders who will be responsible for process development and implementation.

Pages could be written on this issue, but an editorial appearing in SME's *Manufacturing Engineering* written by William F. Hurtubise, Vice President of Engineering, Comau Productivity Systems, summarizes the problems succinctly. This editorial reads:

"In industry's rush to build world-class products with world-class manufacturing technologies, simultaneous engineering became the most overworked and misunderstood term of the '80's, a collection of excuses, false expectations, and marketing lingo. Unless US manufacturers become more enlightened and forthright in using it, they will lose a valuable technique.

"My first exposure to simultaneous engineering was through Fiat's Fully Integrated Robotic Engine (FIRE) 1000 Project. During this four-year project, we achieved some significant objectives, including a 30% reduction in components, 25% reduction in weight, and 50% reduction in manufacturing time. All of them came about because of simultaneous engineering. The result was a world-class engine built by a world-class manufacturing system.

"The marriage of design and manufacturing engineering is the cornerstone of simultaneous engineering's success. Bringing the system supplier into the process provides additional expertise and further assures full integration of the product and the manufacturing system. This is an end in itself, and the product—manufactured with superior quality at the lowest cost—is the primary beneficiary.

"Simultaneous engineering should not be an excuse to form partnerships, although partnerships are necessary for communication. Perhaps the term *partnership* takes on new meaning in simultaneous engineering in that today's system supplier must be more than a good machine or system builder. *Because manufacturers have cut back on their internal manufacturing engineering capability to reduce operation expenses, they often look to machine and system builders to pick up the slack as simultaneous engineering partners.* In fact, under the guise of simultaneous engineering, they may expect the system suppliers to absorb the total cost. This is not simultaneous engineering.

"Another misconception of simultaneous engineering stems from inadequate cost estimating and schedule preparation. As assembly systems become more complex, so do estimating costs and delivery schedules. Normal delivery times do not necessarily index accordingly. Also, integrating product design and the manufacturing process increases the number of product changes that take place. Suppliers may believe that if they get involved earlier in the process, they can help reduce project times and costs. This is a myth. Operating costs tend to go up with simultaneous engineering.

"For the supplier, being involved up-front in the decision making process gives more time to consider alternatives and—with more complete information—to develop accurate prices. Again, more accurate prices do not necessarily mean a budget reduction. When suppliers compete for initial selection, they often base their prices on sketch information. As conditions become more clear, costs may well rise. As for in-process changes, there will be more than ever. With the early start provided by simultaneous engineering, product and process are even more unstable, especially given the understandably competitive price and concept the supplier put forward initially.

"Both manufacturers and system suppliers are trapped to some degree by the sequence of events that lead to a project budget. These events tie back to the initial fund allocation process. The information available then is less reliable than at any other time. Furthermore, in the heat of competition, optimism often outweighs good judgment. The resulting budgets are invariably seeded with time bombs. The bombs go off, one by one, as the project reaches various milestones, leaving gaping holes. The repair procedure is painful at best.

"Whatever happens, however, don't discard the rules used to justify the project. When changes occur, analyze their costs by the justification rules. This procedure weeds out capricious and arbitrary changes, leaving only those that can be justified.

"With the additional time normally available in a simultaneous engineering project, you can explore alternative processes and/or equipment and analyze flexibility issues thoroughly. Perhaps the cost of the manufacturing system can be reduced as a result of this extraordinary analysis. This is certainly a worthwhile objective and a real bonus if it can be achieved. But it is not an objective of simultaneous engineering. Extra manufacturing engineering talent and sufficient time to use it is.

"Simultaneous engineering cannot and should not be an overworked, misunderstood buzzword. It's time for all parties to arrive at a common understanding and change the way they do business with one another. They should analyze how projects are justified, funded, and carried out. True simultaneous engineering is

a stepping stone toward world-class manufactured products. Let's modify the process where needed and take advantage of the opportunities it offers."

Product Development. Attempts to merge product and process development simultaneously is not a new one. In the 1960's and 1970's many corporations had in their product development area advanced manufacturing engineering personnel who were to interact in process selection with the product designer. Unfortunately, in most instances advanced manufacturing engineers did not have procurement authority and often acted without any consultation with the potential operating divisions. Severe arguments between advanced manufacturing engineering people who often designed a product around an emerging or untried process and the manufacturing engineers responsible for system procurement and operation led many companies to abandon this first approach. Simultaneous engineering assumes that there will be a total team activity at the very first phases of product concept, but lack of effective partnership agreements with outside suppliers at this early time remains a major challenge. (Figure 13-5.)

Fiscal Issues. Not the least of the difficulties in trying to establish partnerships is major disparity between the financial resources of the end user and the limited and overtaxed resources of assembly system builders. European and Asian companies have long recognized the need for fiscal aid to outside suppliers while American companies often persist in treating system development as some form of commodity purchase. Fortunately, the increasing number of European and Asian owned companies operating in the United States is forcing American companies to reconsider payment terms and related issues.

Selection of Suppliers. It is becoming obvious that full implementation of simultaneous engineering can force major consumers to develop standards by which suppliers can be engaged or placed on retainer in the same fashion as accountants, law firms and brokers are engaged. This would allow assembly system builders to

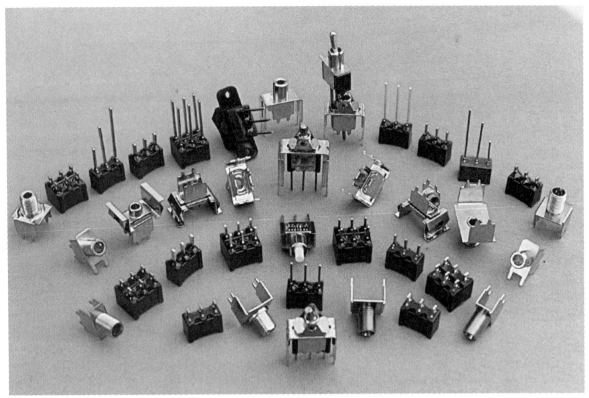

Fig. 13-5. Full cooperation of product design, marketing, component vendors and manufacturing engineers permitted an entire family of switches to be assembled on a single assembly system.

participate from the initial days of product concept. It is an axiom that most of the significant process decisions in any emerging product occur within the first two weeks of concept development. It is extremely rare that any assembly system builder is engaged in these early days. How to handle this issue will be one of the major implementation problems for simultaneous engineering.

SYSTEMS INTEGRATION

One emerging and perhaps irreversible trend is the tremendously expanded obligations and expectations placed on or with assembly machine builders. Until quite recently, assembly machines were for the most part free standing systems which, when shipped to the user facility, would be integrated into shop operations by the engineers and skilled trades of the customer. Even when multiple machines were required to

complete a product, the integration, conveyorization and data collection systems were usually developed by the end user. Increasingly, assembly machine builders are being turned into major system integrators in which they manufacture less of the total overall system and must accept fiscal risk and operating responsibility for major purchased sub-systems such as lasers, welders, machine vision system and data collection systems. The fiscal implications for the systems builder are enormous for such builders. Many of these sub-systems must be purchased several months before machine tryout and shipment, and most will require further development once placed into the initial production phases. Such rapid development of responsibilities places strains on the historical accounting practices used by systems builders, placing higher and higher overhead burdens on their own internal manufacturing operations. This forces many to convert overhead costs to new profit centers. The builders, therefore, are faced not only with developing new processes and equipment but building different forms of internal management skills. There have been at least two major collapses in recent years where customers pushed a chosen system vendor far beyond their existing financial and administrative capabilities to the point of complete collapse. Today's system procurement activities seem almost determined to cause this problem to occur again and frequently. Companies which traditionally looked at their assembly machinery, modular designs and proven station hardware to carry a great deal of the risk of assembly system development and provide the majority of the builder operating profit now find themselves selling process research, integration software and attempting to commercialize emerging technology with a significant exposure to risk and fiscal loss.

The fact that most assembly systems purchased today are multi-machine systems with significant material handling interface requirement has led to several technical developments among major suppliers.

The Trend to Work Cell Plant Design

Increasingly, assembly system builders are expected to interact and directly interface with fabricating equipment producers and eject directly to packaging or cartoning machinery. Figure 13-6 shows a modern video cassette plant in which molding machinery and assembly machinery and packaging machines are coupled with conveyors and storage accumulators. In this facility molding power is transferred to packaged cassettes in less than twenty-five minutes without any human intervention or manual material handling activities. Such plants with proper equipment selection and successful interaction between fabrication, material handling assembly and packaging equipment manufacturers are capable of extremely high sustained rates of production. The system illustrated produces over one-hundred-thousand cassettes every twenty-four hours. It operates twenty-four hours a day, seven days a week, year in and year out. Its net production level exceeds 97% of the total gross production capability. Tremendous production achievements are possible in such work cell plant design but depend on total team collaboration between the various equipment builders.

Fig. 13-6. A work-cell-configured VHS assembly line where conveyors and buffer storage systems coupled molding, assembly, testing and packaging.

New Machine Chassis Developments

The trend toward customers' demands for total assembly and testing systems integration is causing many builders to rapidly expand their line of building blocks or modular indexing, transfer and joining units. Builders formally associated with one or two specific machine configurations now may offer a full range of synchronized, rotary and power-and-free systems together with a significant range of interfacing conveyors and palletizing and depalletizing systems. Many of these builders have developed material handling systems primarily to reduce interface problems and ensure a continuity in the overall system design. The most successful builders are not only developing totally new systems but reorganizing their existing modular construction in a variety of ways. Customers who as-

Fig. 13-7. An integrated automotive product assembly line integrates assembly, test, buffer storage, palletization and production data systems.

sume they know the capacity of specific builders are well advised to visit former and present suppliers to see the degree of flexibility being built into newer modular construction.

The Drive to Direct Integration

Demands for overall manufacturing system design are leading to continuing technological developments and a better understanding of customer expectations in three areas: line balancing, the use of interconnect conveyors and buffer storage and palletization techniques. (See Figure 13-7.)

Line Balancing. The disparities between the production rates attainable with fabricating equipment, assembly systems, gauging and testing systems and packaging equipment precludes any simple conveyorization from one type of a manufacturing system to another. Such line balancing problems have in turn caused a significant amount of industrial research and significant tooling developments to address significant line balancing situations.

Conveyors and Buffer Storage. While line balancing tries to address the disparities of production rates in a line where all systems are operating, there is a further problem of unexpected downtime or stoppages in any one unit of the overall production line. There must be a demonstrated ability to bank or distribute a sufficient quantity of components, partial assemblies or subassemblies between each element of fabrication, assembly, testing and packaging. Such buffer storage not only requires a variety of equipment configurations (hardware) but a tremendous amount of software and research into probable levels of productivity (uptime) maintenance stops establishing realistic norms of expected downtime between failure, and off line storage capability when there is a major downtime problem.

The cost of such integration hardware and software can be a very large portion of a fully integrated production system. In the video tape assembly line shown above *the cost of such integration conveyorization and buffer storage systems was higher than the cost of the assembly system units.*

Fig. 13-8. Modern palletization systems are replacing bulk feeding systems.

TABLE 13-2

KNOWLEDGE BASES

Upper Level Systems

- Artificial intelligence applications to production systems management including process and quality control and inspection
- Integration of CAD/CAM/CAT and its optimization
- Cell controllers and their communications protocol including real time I.F. intelligent servo systems and real time parallel processing
- Technology for acquisition of knowledges for learning control and cooperative processes and their utilization
- Systematization of programming environments for standard programmable devices
- Architecture and construction of autonomous-distributed micro Intelligent Manufacturing System

  Setting Up Processes for Production

- Systematization and organization of setting up technologies including modeling, know-hows, management and evaluations techniques
- Systematization of production scheduling knowledges for systems management

MACHINING TECHNOLOGIES AND MACHINE TOOLS

- Development of self-organizing machining cells and intelligent work systems
- Development of intelligent machining systems for high quality production including quality control techniques
- Indexing and utilization of optimum cutting conditions using learning process control technologies
- Development of intelligent man-machine interfaces
- Development of multi-function, multi-purpose, high reliability processing equipments using new materials
- Advanced tooling technologies using intelligent tools
- Sensor fusion technologies for processing under extreme conditions (e.g. high temperatures)

ASSEMBLY TECHNOLOGIES AND EQUIPMENTS

- Development of autonomous-decentralized assembly robots and cells including flexible tooling, end-effectors, sensors, self-repairing and organizing robots and cells
- Development of an integrated processing/assembly cell applying lithography technique
- Development of work scheduling, process design and servo control systems for product-model based assembly robots
- Systematization and organization of knowledges for product design aiming at improving assembly ability
- Development of intelligent modules for parts supplying, alignment and setting up

MEASURING AND INSPECTION SYSTEMS

- Development of an integrated flexible measuring system
- Systematization of knowledges for the utilization of intelligent 3-D measuring machines
- Recognition technology for 3-D objects and their appearances using artificial intelligence
- Development of a product-model-based automatic measuring machine capable of assessing product's appearance and performance
- Intelligent and mobile general purpose sensing systems

INTELLIGENT PRODUCT DISTRIBUTIONS AND MATERIALS HANDLING SYSTEMS

- Development of advanced metamorphic intelligent AGV with partial capabilities for assembling and inspection including autonomous navigation and map generation systems

TABLE 13-2 (*continued*)

- Acquisition of knowledge and construction of data bases upon scheduling, control and management of materials handling and product distributions including cooperative operations and load balancing among various materials handling equipment
- Development of a metamorphic warehouse capable of flexibly transforming itself e.g. storage space and picking methods to accommodate varying size and shape of the objects to be stored
- Improved international distribution system and standardization of its informations processing

SIMULATION OF PRODUCTION SYSTEMS

- Development of simulation techniques for integrated autonomous production systems using object-oriented languages including computational manufacturing
- Systematization and organization of knowledges on application of production systems modeling and simulators
- Virtual factory based upon integrated autonomous production system

EXISTING PRODUCTION SYSTEMS

- Techniques for standardization of knowledges pertaining to the existing production systems including tools for knowledge codification and systems architecture, data base of systems components
- Systematization and codification to CAI of production systems technologies
- Systems evaluation for integrated CIM i.e. evaluation of performance e.g. economy, reliability, safety, etc. of future integrated CIM systems expected to be realized within 4 to 5 years
- Systems integration techniques based on product model e.g. virtual factories
- Systems integration techniques based on factory model

FUNDAMENTAL TECHNOLOGIES FOR AUTONOMOUS DISTRIBUTED SYSTEMS

- Research into the mechanism of autonomous distributed systems occurring in the nature including bionic systems, biological control
- Application of DNA (including genetic and self organizing) type informations processing mode to production systems
- Control of self organizing production systems including concurrent and localized production systems
- Application of bionic informations transmission mode onto production systems

APPLICATIONS OF NEW MATERIALS

- Systematization and codification of next generation functional materials potentially applicable to the component devices for IMS. Those include self recuperating, coloring, decaying transforming materials

DESIGN TECHNOLOGIES

- Systematization of enabling technologies for products design including organization of basic knowledges, CAD/CAM and data exchange techniques, integrated data bases and development of design data bases
- Systematization and development of process design and production management technologies including organization of current knowledges and application to next generation autonomous distributed production systems
- Development of production activities model based upon product modeling

SYSTEM MODELING

- Modular manufacturing including modules and model-tuning
- Robotic factory including modular robots
- Intelligent factories
- Bionic factories including bionic modules

TABLE 13-2 (continued)

- Biological self-organization including self-organizing modules and cells
- Community type factory including autonomous machinery and holonic module systems
- Free topology system, free topology modules
- Human oriented manufacturing system (HOMS)
- Bottom-up top-down harmonized system
- Ecological harmonic system
- Harmonic type autonomous distributed system
- Cooperative control system
- Advanced autonomous production system
- Unit-lot production system
- Large scale multi-products production

SPECIFIC PRODUCTION SYSTEMS

- Factory capable of continuously producing new products
- Next generation high precision production line
- Holonic production system for machine tools
- Processing and assembling technologies for electronic devices
- Production system for fine chemicals plant

MANAGEMENT INFORMATIONS SYSTEMS

- Intelligent logistics systems
- Systematization and organization of management informations systems (including marketing, personnel management, financing and planning) and their projection towards next generation production systems
- Organization of support systems for R & D activities (planning, data acquisition, computation and monitoring) and their extension towards next generation production systems

HUMAN FACTORS IN PRODUCTION SYSTEMS

- Research into systematization and organization of the quintessence on working amenity in manufacturing environment including its evaluation and monitoring techniques and its sex, age and ethnic group dependency
- Research into human behaviors and responses (e.g. pleasure, fatigue, creativity, sensitivity, etc.) within production systems
- Research and development into human oriented interfaces including their response to human sensitivity and human (especially elderly) support systems
- Research into autonomous man-machine systems and their design
- Standardization of specifications for clean factories

COMMUNICATIONS AND NETWORKING SYSTEMS

- Organization and systematization of manufacturing data, compression/transmission and networking technologies
- Standardization of communications protocol and interfacing for autonomous distributed production systems

Palletization. Palletization between various steps in the fabricating assembly and testing processes achieves a degree of line balancing through the use of a variety of X-Y coordinate storage devices or free floating storage devices carrying components or partially or fully assembled products. The purpose of palletization is, for the most part, not that of line balancing although this is usually achieved through the palletization effort. The main function, however, is providing ability to handle components or sub-assemblies without sufficient physical integrity to withstand damage from traditional material handling processes. Palletization is found extremely often in the manufacture of matched assemblies or in products requiring significant gauging or testing operations between various assembly modules. Growth in the utilization of palletizing is enormous since it is particularly adaptable to work cell configured plants. The technical problems here involve the wide variety of container or pallet configurations and the exposure to damage of the pallets or free floating work-holding fixtures as they are used time and time again in the operating system. It is to be expected that large assembly system builders will prefer to develop their own palletizing and depalletizing systems as they grow ever larger in the overall system integration business. (Figure 13-8.)

EMERGING TECHNOLOGIES

Two major studies have looked forward to determine what areas of technological development were of greatest interest to manufacturing concerns and their management. One was done by the Society of Manufacturing Engineers in their "Profile 21" study. The other study was an initial attempt by Japanese partners in the original Intelligent Manufacturing System Initiative to determine what areas of applied research would prove most useful to Japanese industry. "Profile 21" focused on those new technologies (Table 13-1) that would appear as we enter the twenty-first century. Table 13-2 illustrates a longer view of Japanese industrialists concerning those types of industrial research and development they expect to find on the factory floor well into the next century. In examining these two tables one will find that a tremendous portion of the overall developments and research have direct application to integrated assembly systems. Those responsible

for forward process planning may find these tables of significant interest.

Quality Verification Tools. The never ending search for better and verifiable product quality has led to increasing incorporation of a variety of measurement tools for specific qualities, dimensions or functional characteristics. Implementation of proven laboratory or bench type tools into automated lines has not been easy. Leak testing, shown in Figure 13-4, an extremely common occurrence, requires a significant period of time for vacuum or pressure stabilization before a specific deterioration rate can be determined. Gauging of ever smaller dimensions and matched assemblies requires transducers that not only can measure microscopic dimensional differences but also have the physical integrity to standard repeated cyclic rates on the production floor. Machine vision has at last become user friendly, significantly reducing the time to program and the skill levels required to program for specific application and more importantly to convert computer images into specific dimensional information. The emergence of rapidly developing automated quality measurement tools occurs on such a frequent basis that prior experience has little value in discussing proposed systems.

SUMMARY

The tools for integrated manufacturing systems exist or are rapidly emerging. The integration skills of major systems builders are constantly growing. The major problem in a simultaneous engineering environment is the establishment of proper contracts or financial arrangements between the system integrators and their customers. The problems of automation implementation remain more in the managerial field than in technological innovation.

Author Index